Oakley Coles, Robert Ramsay

# The Mechanical Treatment of Deformities of the Mouth,

## Congenital and Accidental

Oakley Coles, Robert Ramsay

**The Mechanical Treatment of Deformities of the Mouth, Congenital and Accidental**

ISBN/EAN: 9783337239770

Printed in Europe, USA, Canada, Australia, Japan

Cover: Foto ©berggeist007 / pixelio.de

More available books at **www.hansebooks.com**

THE

# MECHANICAL TREATMENT

OF

# DEFORMITIES OF THE MOUTH,

## CONGENITAL AND ACCIDENTAL.

BY

### ROBERT RAMSAY,

MEMBER OF THE ODONTOLOGICAL SOCIETY;
AUTHOR OF TWO PAPERS ON THE TREATMENT OF CONGENITAL CLEFT PALATE, ETC.

AND

### JAMES OAKLEY COLES,

HONORARY DENTIST TO THE HOSPITAL FOR DISEASES OF THE THROAT, GOLDEN SQUARE;
MEMBER OF THE ODONTOLOGICAL SOCIETY, ETC.

LONDON:
JOHN CHURCHILL & SONS, NEW BURLINGTON STREET.
1868.

TO

# SIR WILLIAM FERGUSSON, BART., F.R.S.,

SERGEANT-SURGEON TO THE QUEEN,

THIS VOLUME IS,

BY PERMISSION, RESPECTFULLY DEDICATED.

# PREFACE.

IN bringing this volume before the members of
the Dental and Medical profession, we desire
to state our reasons for thinking that such a
work would not be unacceptable. For some few
years past the treatment of Congenital Cleft Palate
has formed a subject of considerable discussion
before both Medical and Dental societies ; and our
journals have always been ready to receive com-
munications on a subject that seemed to command
so much attention.

The recent advances that have been made in the
mechanical treatment of this deformity have not
lessened the interest that had already arisen ; and
as no publication on the mode of procedure and
account of cases had appeared since the time of
Snell, in 1828,—apart from reprints of papers
read before the learned societies,—we were in-
duced to think that the present book, though by
no means so complete as the subject deserves,
would be of interest to those gentlemen who have
devoted time and attention to this speciality, as
well as to the general practitioner.

Where we have given descriptions of the *modus
operandi* of any part of the treatment put forward,

we trust we shall have been found to be sufficiently explicit; and in the chapter on the appliances used in the past, and at the present time by other gentlemen, we hope it will be recognized that we have endeavoured to be impartial.

In reference to the account of six cases, treated by mechanical means, of loss of parts arising from accidental causes, we feel some explanation is necessary. The novelty in these instances does not consist so much in the general line of treatment adopted, as in the fact that where elastic rubber has been used it has been adapted in metallic moulds, made expressly for each case, to the outline of the deficiency which it was intended to supply;. the facility with which this can be done being considerably increased since the first manufacture of elastic rubber in this country, about eighteen months since, so pure and so carefully prepared as to be suitable for dental purposes. Before this date we were dependent on America for our supply.

We have to thank many medical and professional brethren for the cases they have placed in our hands for treatment; but there are obvious reasons for our not mentioning their names in those instances where we have given a report of the treatment pursued.

68, WIMPOLE STREET, CAVENDISH SQUARE, W.,
*October,* 1868.

# CONTENTS.

## CHAPTER I.

## CHAPTER II.

## CHAPTER III.

## CHAPTER IV.

## CHAPTER V.

# CHAPTER VI.

# CHAPTER VII.

# CHAPTER VIII.

# DEFORMITIES OF THE MOUTH.

## CHAPTER I.

### ON THE ORIGIN AND DEVELOPMENT OF CLEFT PALATE.

THE predisposing causes of cleft palate have often been a matter of careful research and reflection; but up to the present time no satisfactory reason has been given for the development of this deformity. It has given rise to great speculation and conjecture, but we cannot affirm that we are any nearer the truth now than we were many years ago. That it is in some instances caused by arrested development, the want of substance in the parts would seem to indicate; but in others it is apparently more from a want of union in the median line at the proper time than from any lack of material to produce the perfect palate, since the margins of the cleft in many cases supply more than enough to fill up the deficiency in the operation of staphyloraphy, while in others again no amount of skill would be able to bring the opposite sides into such contact as to get permanent union. This variety would then appear to point to the fact that while in some cases the lesion may occur from arrested development during the whole time of pregnancy, in others it occurs only at that time when the parts should unite in the central portions, this condition being consequent, according to Dr. Engel's opinion, to increased breadth of the anterior portion of the head, which is again caused by a variety of conditions in embryonic life, such as congenital hernia cerebri, dropsy of the third ventricle, or of the anterior cornua of the lateral ventricles, or excessive development of the anterior cerebral lobes.

This view of the case becomes the more intelligible if we refer to the statements of embryologists as to the condition of the embryo about the thirty-eighth day.

The annexed woodcut, copied from the work of M. Coste, shows the manner in which the parts ultimately producing the cavity of the mouth and the adjacent structures are developed.

The discovery of certain processes in the human embryo, termed visceral or branchial arches, was first made by Rathke, who saw a fancied resemblance to the branchial arches of the fish, and therefore gave the latter of the two names mentioned. At first these segments are incomplete, originating simply in buds on each side of the anterior portion of the embryo ; but as the growth proceeds they become elongated into processes, which, ultimately meeting in the median line, produce what Rathke has called visceral arches. The superior maxillary bones, the lower jaw, and

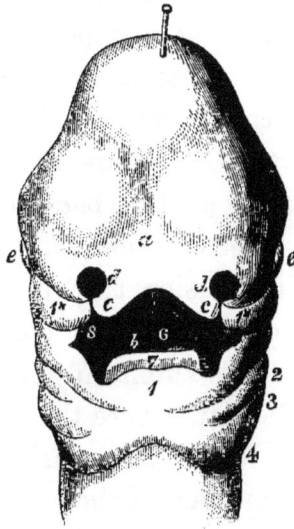

Fig. 1.

Mouth of the Embryo at the 38th day.

| | |
|---|---|
| *a*. Median bud. | 1\*. Superior maxillary bones. |
| *c c*. Incisor buds. | 2, 3, and 4. Second, third, and |
| *d d*. Nostrils. | fourth visceral arches. |
| *e e*. The eyes. | 6. Septum of the nose. |
| *b*. The mouth. | 7. The tongue. |
| 1. The lower jaw. | 8. Roof of the mouth. |

two of the bones of the ear are developed from the first of these arches, the two halves of which are separated by a space immediately below the frontal eminence.

The jaws being thus developed in two segments, meeting in the median portion under natural circumstances, and the central portion of the lip being developed from a separate part to that of each side, we can readily understand how the arrest of development of any of these parts for ever so brief a time at this period of embryonic life would lead to great deformity. Cleft palate may indirectly throw some light upon the physiology of this subject, if we remember that the cleft is invariably at the back of the palate, except in cases of association with hare-lip ; thus leading to the conclusion that union of the two halves of the upper jaw takes place from before backwards, and not uniformly in their entire length ; for if the latter were the case, we might naturally expect to find an occasional instance of perforation of some part of the palate in the region lying between the alveolar ridge and the apex of the uvula; but we are not aware of any such case having ever been seen.

In consequence of the distension within the cranium of the embryo, the parts on each side of the palatal fissure appear to be not only deficient in the median line, but more widely separated than under the natural condition they would be ; the distance between certain parts, such as the infra-orbital foramina and nasal processes of the upper maxillary bones, being taken as points for comparison with measurements from the same parts in the healthy new-born child.

Dr. John Smith, of Edinburgh, brought this matter prominently before the scientific world by reading a paper on " Certain Points in the Morphology of Cleft Palate " at a meeting of the Royal Society of Edinburgh, and gave to the subject additional interest by his philosophical reflections on the connection between the measurements taken at birth and those obtained from the mouths of adult cases. Having taken the first bicuspid of each side as a point of measurement (as the least change takes place here from infancy to adult life), in the normal dental arch averaging from one and an eighth to one and a quarter of an inch, Dr. Smith gives the measurement of sixteen cases of congenital cleft palate, in the full-grown subject, which show an average width somewhat less than in the perfect formed jaw.

" In six cases where the intermaxillary bones seemed altogether absent—probably instances of double cleft where they had been removed by the surgeon, or where they had never been developed—

1 case measured $\frac{3}{8}$ of an inch,
1        ,,          $\frac{5}{8}$    ,,
2        ,,          $\frac{7}{8}$    ,,
1        ,,          1 inch,
1        ,,          $1\frac{1}{8}$ of an inch,

giving an average measurement of between $\frac{6}{8}$ and $\frac{7}{8}$ of an inch.

" In ten cases of simple cleft palate alone, or of cleft palate combined with only unilateral fissure,

1 case measured $\frac{5}{8}$ to $\frac{6}{8}$ of an inch,

1      ,,      $\frac{5}{8}$ of an inch,

4      ,,      $\frac{7}{8}$      ,,

3      ,,      1 inch,

1      ,,      $1\frac{1}{8}$ of an inch,

giving an average measurement of $\frac{7}{8}$ of an inch.''

Dr. Smith then goes on to say that it would thus appear that while in the infant there is abnormal separation, in the adult there occurs abnormal approximation of the parts on each side of the fissure, as in fig. 2. To a certain extent this

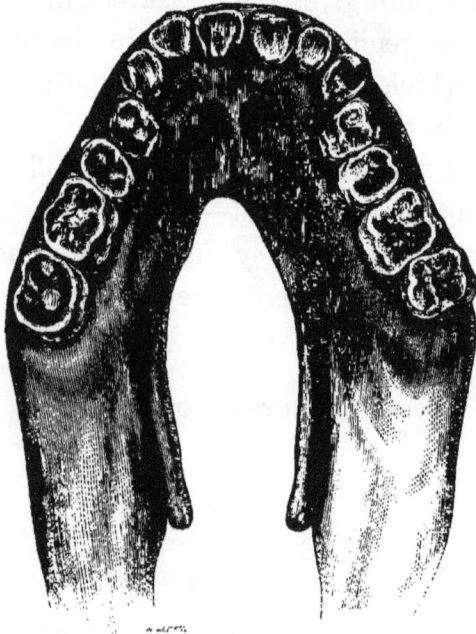

FIG. 2.

approximation of parts may be fortuitous—a mis-direction of growth, dependent upon the absence of the mesial structures, while the superior maxilla is becoming, as age advances, elongated downwards by the expansion of the antrum.

But as the same approximation seems to occur even where only a partial fissure exists—the cleft being limited to the palate, while the maxillary arch is throughout complete—there is reason to conclude that it is in some measure to be considered as an effort on the part of nature towards reparation, or rather amelioration of the existing defect.

We had for some time previous to the reading of Dr. Smith's paper been attracted to the subject of which it treats in the latter portion, by a number of cases giving most marked evidence of this approximation in the region of the first and second bicuspids, though we had not thought to apply to it the idea of its being a natural effort to decrease the *size* of the cleft.

We came then to the conclusion, and have had no reason up to the present time to doubt its soundness, that it was rather owing to the eruption of the molars and the pressure they exerted than to any movement originating *per se* in the portion of the jaw indicated by the position of the bicuspids.

If the upper jaw be taken as representing an arch, it will, we think, be conceded that where the bicuspids are developed would be the weakest part of that arch, and supposing the central portion to be well developed and of a symmetrical shape, any pressure applied from behind forwards would produce irregularity. With a perfect palate such pressure will in the upper jaw very often produce a bulging out of the bicuspids; in the cleft palate, the median portion (which in the natural state ties the two piers together) being absent, the molars are to some extent separated, while the first bicuspids are approximated.

As curiously corroborative of this view of the matter, we would submit the instances of irregularity that occur in the lower jaw, where we have a condition somewhat analogous to the cleft palate and imperfect arch of the upper jaw. In consequence of the absence of any strong tie between the two angles of the dental arch in the inferior maxilla, it is often seen when the wisdom teeth are erupted, and all the other teeth are perfect and in good position, that the bicuspids will in some rare cases sink below the level of the rest of the lower teeth, in consequence of the absorption produced by the pressure from behind forwards, and in many cases an approximation of them takes place towards each other identical with that which we have brought forward as occurring from a similar cause in the upper jaw.

This is further corroborated by the fact, that in the first dentition, in the simple cleft there is no deformity of the dental arch. This is shown in fig. 3, the model of a child's mouth at four years of age, and drawings of several other casts could be shown in proof of the same fact.

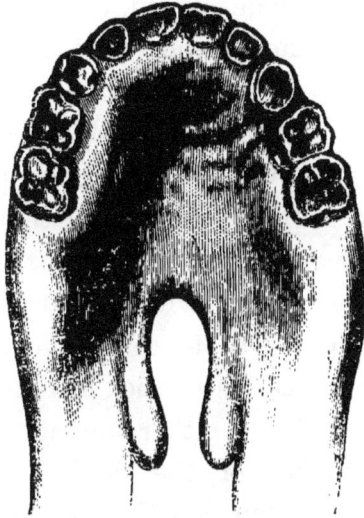

FIG. 3.

On these grounds, therefore, we would, with all deference, submit this view of the case to our professional readers.

Whether cleft palate is hereditary or not it seems impossible to determine, though from the evidence we have in hand we should come to the conclusion that it is not. Still, unless one could obtain accurate records of ancestry for three or four degrees of removal, it would be impossible to assert anything with certainty. We feel, however, that it may confidently be asserted that this deformity cannot be produced from any impression received by the mother during pregnancy, for in every case which has come immediately under our notice, where one of the parents has had cleft palate, all the children born have been perfectly developed, even though dread on the part of the mother of transmitting the deformity has been most constant.

In one case, curiously enough, there are three members of one family with cleft palate, one seventeen years of age, the other thirty, and the third thirty-five; the first and last are ladies, the other a gentleman, who is married, and has a family without any trace of the father's deformity. In these cases no instances of cleft palate could be found either among ancestors or collateral branches of the family. It will be interesting to watch whether in the following generations any traces spring up again, for no cases of immediate transmission seem to have been placed on record.

It is strange, and, so far as we are aware, quite unaccounted for, that the majority of cases are of fair complexion and nervous temperament, while very few are of a dark complexion and lymphatic temperament. Whether this conclusion will be justified by a larger field of observation we cannot of course say; but in a great number of instances we have found it to be so.

We have always deemed it a matter of great comfort to parents to be able to assure them that there is little liability of their children inheriting their deformity; and as yet we have seen no reason to render this assurance unreliable.

# CHAPTER II.

## ON THE ANATOMY AND PHYSIOLOGY OF CLEFT PALATE.

For the anatomy of cleft palate we are almost entirely indebted to Sir William Fergusson, who some years back had the rare good-fortune to come across a case in the dissecting-room, an account of which he gave in a paper read before the Medical and Chirurgical Society on the 10th December, 1844. On the conclusions which he came to as to the physiology of the parts he based his method of treatment for this condition of the palate, and put forward the plan of dividing the levator palati muscle, in order to obtain perfect control over the palate during the operation.

The value of this account of the anatomy and physiology of cleft palate cannot be overestimated, since, in addition to the light it threw upon the surgeon's work, it has of late years become the basis of treatment by mechanical means.

Under these circumstances we feel we cannot do better than give to our readers an extract from the Society's Transactions (for 1845), in Sir William Fergusson's own words :—

" Few have had the opportunity of dissecting a cleft palate, and some notice of a specimen in my possession will form an appropriate introduction to the views developed in this paper. The fissure in this instance implicates a portion of the hard as well as the whole soft palate, and is such as the surgeon frequently meets with in practice. The specimen was procured in the dissecting-room from the mouth of an aged female subject.

" In the examination of this preparation there are several marked differences between it and the parts in a more natural state. The superior constrictor muscle is more fully developed than under ordinary circumstances, and its upper margin, extending between the basilar process of the occipital bone and the internal pterygoid plate is particularly distinct. This part of the muscle forms a sort of semicircular loop, in which the levator palati muscle seems to be suspended.

"The pharynx has been laid open by a perpendicular incision through the constrictors in the mesial line, and the movable portion of the palate has been dissected on one side. The circumflexus, or tensor palati, differs little from the natural condition, and the levator palati is much as it is usually met with, its lower end spreading out in all directions on the soft palate. The palato-pharyngeus consists of two distinct bundles

# DESCRIPTION OF COLOURED PLATE,

Showing a Dissection of a Cleft Palate; copied from the
work on Practical Surgery, by the kind permission of
Sir William Fergusson, Bart.

---

The plate represents the posterior nares and upper surface of
the soft palate.

a. The levator palati; the dark line shows where it should
be cut across.

b. The inner bundle of fibres of the palato-pharyngeus
forming the posterior pillar of the fauces; the black
line indicates the place for division.

c. The palato-glossus, with the mark for incision, if one
should be deemed necessary.
The tonsil lies between these two muscles.

d. The tensor palati, the cartilaginous extremity of the
Eustachian tube is in front of this letter.

e. The posterior extremity of the inferior turbinated bone.

f. The septum.

g, g. The uvula on each side stretched apart.

of fibres; one, the smaller of the two, running
between the tensor and levator palati; the other,
a mass equal in size to a goose-quill, seems to
form the principal part of the free portion of the
palate; and posteriorly its fibres, previous to
joining those of the other bundle, form the
whole muscular portion of the posterior pillar
of the fauces. This muscle arises by tendinous
and fleshy fibres from the posterior margin of the
osseous palate and the inner surface of the internal pterygoid plate, and takes its usual course
and attachment posteriorly. A bundle of fibres,
about the size of a crow-quill, can be traced along
the lower border of the inner margin of the soft
flap. These fibres extend between the posterior
margin of the hard palate and the uvula, and are
probably analogous to the azygos uvulæ. The
palato-glossus can scarcely be distinguished. A
small arterial twig, doubtless a branch of the
ascending pharyngeal artery, can be traced between the levator and tensor palati muscles.
The throat and upper part of the pharynx
generally is smaller than in the well-formed
state, but the deficiency in the mesial line of the
palate seems more the result of a want of union
than of the usual materials of the velum (see
page 9).

"The act of deglutition in the natural state of
the parts, while food is passing through the upper
end of the pharynx, has been a subject of

considerable speculation among physiologists, especially with reference to the manner in which the communication betwixt that bag and the posterior nares is closed for the time being.

"It has been pointed out by Dzondi and Müller that the palato-pharyngei muscles, when fixed in the soft tissues at their upper ends—as in the natural state of the velum—must, during contraction, tend towards the mesial line, and so by their approximation diminish the capacity of the throat. But in the cleft state there is no central fixed line, and each muscle, acting between its extreme attachments—viz., the palatine bones above and the thyroid cartilage below— must, during contraction, tend to widen the throat rather than close it. In the condition alluded to, these muscles, joined with the levatores palati, have the effect of enlarging the gap in the mesial line. It is evident that the doctrine of the above-named physiologists will not account for the closing of the aperture under these circumstances, and how then is the occlusion effected? I am not aware that it has ever been accounted for. Malgaigne,* in describing the simple fissure of the palate, has alluded to the approximation of the edges during deglutition, ' by a muscular action,' as he says, ' of which it is difficult to give an explanation.' I think that any one who looks at the preparation in my pos-

* "Manuel de Médecine Opératoire," Paris, 1834, p. 486.

session can have no doubt as to this movement. The superior constrictor has evidently the power of throwing the two lateral portions of the palate forwards and inwards, so that they are forced into contact in the mesial line, and thus the back of the fissure is closed, while the constrictor is acting on the upper part of the pharynx, like a broad semicircular band. The upper border of this muscle, as it is seen in the preparation alluded to, must evidently have the effect described, and the lower fibres will act still more effectually, in consequence of there being no connection mesially to prevent them starting forwards during contraction, so as to stretch across, almost in a direct line, extending between the lateral attachments of each muscle. Some of the fibres of the middle constrictor may also aid in this movement. The palato-pharyngei muscles are thus forced into contact, and their ends, behind and below the parts so held in apposition, may then act in the manner described by Müller, while possibly the thickness of the two portions of the soft palate may be increased by the contraction of each palato-pharyngeal muscle at the points of contact. The azygos uvulæ may probably contribute to the latter effect. . . .

" As we look into the open mouth, the flaps may be seen under four different conditions. First. If the parts be not irritated in any way, the gap will be quite conspicuous, the lateral flaps will be distinct, and the posterior nares, with the upper end of the pharynx, will be observed above and behind. Second. If the flaps be touched, they will in all probability be jerked upwards by a motion seemingly commencing at the middle of each. Third. If the parts be further irritated, as by pushing the finger against them into the fissure, each flap is forcibly drawn upwards and outwards, and can scarcely be distinguished from the rest of the parts, forming the sides of the nostrils and throat. And, fourth. If the parts further back be irritated, as in the second act of deglutition, the margins of the fissure are forced together, by the action of the superior constrictor muscle, already described in my observations on this process, in an earlier part of the paper.

"All these conditions and movements are, in my opinion, very readily accounted for. In the first instance the parts may be deemed in a quiescent state; in the second, the levatores palati are called into play, and move the flaps as described; and in the third, these muscles act still more forcibly, and the palato-pharyngei will join in drawing the parts outwards. The fourth condition I need not again describe.

"If the free margin on one side of the fissure be seized with the forceps, drawn towards the mesial line, and the flap be then irritated, it will be drawn upwards and outwards with remarkable force; this movement, it is evident, can only be effected by two muscles, the levator palati and palato-pharyngeus. These muscles, then, I consider the chief mechanical obstacles to the junction of the margins in the mesial line. Hitherto I have taken no notice of the action of the circumflexus, or tensor palati. I am inclined to think that its action is very limited, and probably, as the dissection in my possession would indicate, is greater upon the parts outside the posterior pillar than on those contiguous to the fissure. Neither have I alluded specially to the action of the palato-glossus, because, though it might with a feeble power incline the soft palate downwards, its influence, as regards the practical view I am now taking, is completely counteracted by the more powerful muscles connected with the palate above."

There can be no doubt that the plan suggested in the concluding portion of this paper by Sir William Fergusson, of dividing the levator palati, palato-pharyngeus, and palato-glossus muscles, is by far the most scientific and certain way of proceeding in order to get an easy approximation of the margins of the cleft; and in the recent lectures on the " Progress of Anatomy and Surgery during the Present Century," delivered at the Royal College of Surgeons, the number of cases (between 300 and 400) which that gentleman has treated clearly show the soundness of the views which he put forward in 1844.

There seems, however, one important point that has been almost entirely overlooked, that is, the deformity which invariably exists above and behind the soft palate, in consequence of which the upper part of the pharynx entirely loses its dome-like form, the ends of the turbinated bones being exposed to view, and the posterior openings to the nares absent.*

When, therefore, the palate which has been subjected to a surgical operation is brought into play, the parts would seem to be perfect, and much dissatisfaction is felt at the disagreeable tone of the voice, often forming a matter of surprise both to patient and operator. There can be little question that in very many of these cases this has arisen from the free communication that exists between the upper part of the pharynx and the cavity of the nose, even when they are separated from the mouth partially or completely by the now perfect velum palati.

All the skill of the surgeon would, we fear, fail to restore the posterior nares to their natural condition, and yet it is tolerably clear that we cannot expect to get a natural tone of voice,

---

* Passévant relates a case in which he attached the uvula (after the cleft had been closed) to the back of the pharynx, in order to improve the voice ; but the practice does not seem to have become at all general.

accompanied by intelligible articulation, unless something can be done either surgically or mechanically to represent the parts that have been undeveloped in the cavity of the nose and pharynx.

When the cleft extends into the hard palate to any extent, the septum of the interior of the nose will be found incomplete, as in fig. 15, page 54; and where the cleft is unsymmetrical, the margin of the gap in the anterior portion will be seen articulating with the vomer very often. Since the propagation of clear and agreeable sound is dependent to such a degree for its modification on the shape of the passages through which it travels, the importance of considering the nasal region in the treatment of congenital cleft palate cannot be overlooked.

# CHAPTER III.

WHEN a child is born with hare-lip, the attention of the medical practitioner or nurse will be at once attracted, and an examination made of the mouth to ascertain whether it is complicated with cleft palate. If, however, there is no deformity of the lips, the simple cleft may not be observed until the child begins suckling; if it is small, and confined to bifurcation of the uvula, this will not cause much trouble; but should it extend through the soft palate into the hard, the milk will be found oozing from the nose instead of passing from the mouth into the stomach in the normal manner.

Under these circumstances, the child will have to depend for its support upon the nourishment that can be administered to it by means of a spoon or feeding-bottle. The latter is the better course fo the two undoubtedly, and the little patient may be very much helped in the process of receiving its food by means of the artificial nipple and tubing attached to Maw's very simple but efficient feeder if a little contrivance shown in the accompanying woodcut be attached to the neck of the mouthpiece.

FIG. 4.      FIG. 5
The palate-piece alone   and   attached to the ordinary nipple
sold with Maw's infant-feeder.

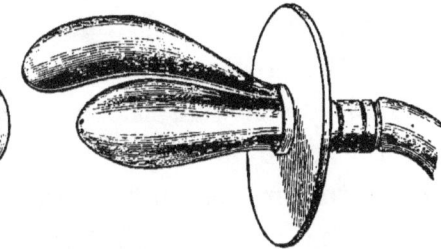

This consists simply of a flap of elastic india-rubber made to fit into the roof of the mouth. The pressure of the nipple against its surface when in position will thus convert it into an artificial palate-piece, and prevent the escape of the milk into the nose in the effort of swallowing. It was suggested some years back that a flap of thin sheet elastic, not modelled to the arch of the palate, but simply cut out and sewn on to the feeder should be used; when it is not convenient to obtain such a one as we have just described this is a very good expedient, but when it is possible to procure a properly-fashioned arrangement, it is evident the discomfort and anxiety that must arise will be very much lessened.

Sponge or leather* is sometimes used for this purpose, but is on many accounts very objectionable, from becoming sour and offensive after use, while the vulcanized rubber can be kept perfectly sweet and wholesome by means of washing in warm water.

Under the most favourable circumstances, and when the greatest possible care is exercised, there is little doubt that the growth and development of the child is very much retarded, though its constitution may not ultimately suffer in consequence of the defective palate, beyond that delicacy of frame which is very often the accompaniment of extreme parental anxiety and watchfulness.

The time for operating upon the hare-lip will have to depend upon the state of health of the babe, and may take place a few hours or several months after birth, according to circumstances. This is a matter exclusively in the hands of the surgeon, and quite beyond our province to enter into, except in reference to one or two points which, with all deference, we would submit to the consideration of those gentlemen who are in the habit of performing the operation.

When the division in the lip is bilateral, and the intermaxillary portion of the jaw very pro-

* Snell, on " Artificial Palates." 1828.

minent apparently, it is usual in a large number of cases to remove the protuberance altogether, using the middle portion of the lip for the columnæ nasi, and then, having pared the edges of the side-flaps, to bring them together in the median line.

The result of this treatment in after-life is to give to the upper lip an exceedingly flat unsightly appearance, and to the lower lip a relaxed and pouting expression.

If the patient's face be looked at from the side, the contour of the countenance would seem to indicate that only a portion and not the whole of the intermaxillary process should have been removed, since its apparent prominence was undoubtedly due principally to the great want of substance on each side, and not alone to an excessive development in the median portion. In Sir William Fergusson's recent work "On the Progress of Anatomy and Surgery during the last Century" this is suggested as a reason, and drawings given of a case, by which it is proved how little deformity need occur when the patient is in the hands of a skilful operator. In the same chapter it is also pointed out that the notch in the lip, so often observed in after-life, is mainly owing to the edges of the divided lip not being cut away to a sufficient extent, so as to bring the skin and mucous membrane into proper contact with each other. We shall have occasion to

C

remark in the sixth chapter of this volume on the trouble arising from the neglect of this apparently trivial matter. During infancy but little difference is observable in the tone of the babe's cry, but when the time arrives at which the child should, under ordinary circumstances, begin to give utterance to "articulate" sounds, it becomes apparent how great a change the deformity gives rise to in the voice.

In many cases no amount of attention, except a mother's instinct, will be able to understand what the child attempts to express; and this, strange to say, is not regulated, as might be expected, by the extent of the cleft, but is almost as bad in simple division of the soft palate as in those cases where there is also division of the hard palate and alveolus.

From birth up to seven or eight years of age the cleft increases both in width and length, in proportion to the rest of the mouth, unless, in the case of hare-lip being associated with it, there are mechanical means used to compress the parts laterally, by the use of a truss, such as we find described in the work by Sir William Fergusson already referred to. The gap may then be reduced so that at the anterior part, the two sides, of the divided alveolus will become so closely approximated as to appear continuous. After the age we have mentioned we are led to the conclusion, so far as our own opportunity of observation has gone, that the cleft simply increases in length, the width being, in the majority of cases, the same at twenty-one years of age as at the earlier period.

From seven to ten years of age the patient will become first conscious and sensitive of his or her defective speech. At this age, too, nourishing food properly masticated is of great importance. Both these circumstances, therefore, would seem to point to it as the best time for the insertion of an artificial velum, unless the operation of staphyloraphy has been performed during infancy;* for it will be well-nigh impossible in a case of cleft in the hard and soft palate to apply the food with the tongue to the roof of the mouth in such a way as to ascertain when it is ready for swallowing. The consequence will therefore be that a considerable quantity of that which should yield the most nutriment is received into the stomach in such a state as to impair the digestive organs. There is also the liability during childhood of the cleft getting filled up with the solid food, and in some cases causing suffocation.

The difficulty of controlling the passage of air through the nares by means of the velum palati

* It should be remembered, however, that this operation has been so recently introduced that we have no means of judging what its effect will be when the patient arrives at adult years.

in nearly all cases gives rise to a most pernicious habit of checking it by means of a contraction of the alæ of the nose through the influence of the compressor naris muscles; thus to a certain degree rendering the utterance more distinct, though it gives a very disagreeable "nasal twang." It is wonderful to what an extent these muscles come under the patient's control. In the French language this would be of no consequence; in the Anglo-Saxon tongue, however, it is a considerable annoyance, and most difficult, when firmly established, to overcome. Still, if an artificial palate be inserted at the time we have indicated, it may be prevented or checked to a very great extent, until the habit is entirely forgotten or overcome by the patient finding it is unnecessary to control the passage of air through the nostrils, except by the means now provided in the elastic velum.

From the great exposure of the nares and fauces to the air in large perforations or clefts of the palate, these parts are exceedingly liable to inflammation and ulceration extending downwards, and during the winter months causing frequent sore-throat and deafness, as well as loss of voice. Considerable irritation arises in the cavity of the nose from the mucus drying on the turbinated bones and margins of the opening, in consequence of the too free access of air to these parts. All of these symptoms, however, disappear when the cause is removed, either by operation or mechanical treatment, though the latter, by checking to a greater extent the passage through the nares, is probably the more efficient of the two for this purpose.

# CHAPTER IV.

SOME ACCOUNT OF THE APPLIANCES USED FOR REME-
DYING CLEFT PALATE (WHETHER CONGENITAL OR
ACCIDENTAL) FROM A.D. 1552 TO THE PRESENT
TIME.

AN account of the progressive stages by which
we have arrived at the present comparative perfection of artificial palates may not be uninter-
esting to our readers, or out of place in a work
of the present kind.

Our principal authority on this subject is Snell,
who took great pains to collect all that it was
possible to glean as to the contrivances used by
our forefathers for remedying this deformity.

Little is known or said on the matter till the
fifteenth century, though Isaac Guillemeau, in his
work published in 1649, mentions the name by
which the Greeks called the appliances for filling
up the cleft; thus leading us to infer that they
were acquainted with some method of treatment
for perforation or cleft of the palate.

In order that we may more easily see the
time that was occupied in passing, stage by stage,
from one improvement to another, we propose to

arrange the names of those surgeons, dentists, and others who have paid any attention to this matter, in chronological order.

1552.—Hollerius, in his "Observ. ad Calcem de Morbis Internis," proposes to stop the apertures with wax or sponge.

1565.—Alexander Petronius, in his "De Margo Gallico," proposes, when there is but one opening in the palate, to stop it with wax, cotton, or a gold plate, taking care to give to the instruments the same concave form as the roof of the mouth. Though this is the first mention of a gold plate being used for this purpose, still, from the fact of Petronius not being more explicit as to its mode of fitting and retention in the mouth, we are, as Snell very justly observes, led to the conclusion that the remedy was one with which his readers were not altogether unacquainted; and we must not therefore give Petronius the credit of being the inventor of this mode of treatment.

1579.—Ambrose Paré, in his book on surgery, published in Paris, and in the year 1649 translated into English by Thomas Johnson, proposes that the cavity should be covered over by a gold or silver plate, "made like unto a dish in figure, and on the upper side, which shall be towards the brain, a little sponge must be fastened, which when it is moistened with the moisture distilling from the brain will become swollen and puffed, so that it will fill the concavity of the palate, that the artificial palate cannot fall down, but stand fast and firm as if it stood of itself."

1649.—Isaac Guillemeau, in his " De Ouvres," gave a drawing of an instrument similar in form to Ambrose Paré's instrument; but suggested that, as it was not always possible to adapt the plate perfectly to the roof of the mouth, a lining of sponge or lint should be applied, in order to render the closure more complete.

1653.—Amatus Lusitanus, in his " Curat. Medic. Centur.," mentions a boy with diseased cranium and perforated palate, whose voice was restored by means of the gold plate and sponge.

1685.—Nic. Tulpius, in his " Observat. Medici," mentions the same mode of treatment.

1715.—Garangeot, in his " Treatise on Instruments," is the first that we find making any step in advance of his predecessors with regard to the construction of obturators. Describing one, he says :—" This instrument has a stem in the form of a screw, upon which runs a nut. To make use of it, take a piece of sponge, cut in the shape of a hemisphere, with a flat surface ; pass the stem of the obturateur through the sponge, and fix it by means of the nut. Dip the sponge in water, squeeze it dry, and introduce it through the aperture."

1723.—Fabricii Hieronimi, in his " Chirurgicis Operationibus," recommends sponge, lint, or silver plate; not suggesting any new form of instrument. He is the first, so far, that is, as we have been able to examine these old works, who makes specific mention of congenital cleft palate in contradistinction to accidental cleft or perforation.

1734.—R. Wiseman, Sergeant-Surgeon to King Charles II., in his Chirurgical Treatises gives evidence of having bestowed much thought upon the treatment of the defects of the palate, though he cannot be said to have made much real and practical progress. His novelty in treatment consisted in filling up the cleft with a paste composed of myrrh, sandarac, and a number of other ingredients. His idea was certainly in advance of his time; for by this means a most important end was gained,—that of perfect exclusion of air by its complete adaptation to the margins of the cleft. We are unfortunately not informed how this "paste palate" was kept in position.

1739.—Heister, in his "Institutions of Surgery," suggests the use of "a gold or silver plate adapted to the perforation, and furnished with a handle or small tube, which, being armed at the top with a sponge, he may thereby exactly close the perforation."

1754.—Astruc, in his "Treatise on Syphilis," makes the first mention that we have of a silver button to the metallic obturator, in place of the sponge, in order to avoid the unpleasantness arising from the absorption of mucus.

1786.—M. Pierre Fouchard, in his "Chirurgeon Dentiste," gives an account of some instruments which show a very great improvement on the forms previously in use; the sponge, as a means of support to the obturator, being substituted by an arrangement of metallic wings, worked into proper position after introduction into the cleft by means of a hollow stem and nut, which, when screwed down, kept the wings (covered with soft sponge) across the aperture.

There are descriptions given of others on the same principle, and of one on a then new plan, depending for its support upon ligatures round the canine teeth.

It will thus be seen that more than two hundred years had elapsed before any decided improvement took place in this department of dental science. MM. Dubois Foucou, Touchard, Bourdet, Cullerier, and De Chamont give descriptions of a variety of obturators, all more or less resembling the instrument of Fouchard, with its arrangements of wings, clasps, and screwnuts.

1820.—The next advance made was by M. De la Barre, who is the first to mention the use of "elastic gum" in the restoration of the velum and uvula. The artificial palates designed by this gentleman were ingenious in the extreme, but of such a complicated nature that none but a man of considerable mechanical genius could ever hope to be successful in their application. Still we must bear in mind the great step taken towards the present instruments in use by the introduction of " elastic gum."

1828.—We now come to a consideration of the artificial palates constructed by Mr. Snell, who arrived at much more satisfactory results in his method of treatment than his predecessors could have done, from the fact that he first obtained an accurate model of the mouth, on which he mounted and fitted his obturator—a point that up to this time is not mentioned, even if it was practised.

He says in his book that, with the exception of one method proposed by Mr. Alcock, in the *Medical Intelligencer*, he is not aware of any successful mode of treatment for remedying congenital cleft of the palate excepting his own plan, which he goes on in the next pages to describe in the following words :—" My method of constructing an obturator is, with a gold plate, accurately fitted to the roof of the mouth, extending backward to the os palati, or extremity of the hard palate, a part of the plate, about an inch in length, being carried through the fissure. To that part of the plate which answers to the nasal fossæ are soldered two plates, meeting in the centre and carried upwards through the fissure to the top of the remaining portion of the vomer, to which it should be exactly adapted, and made to the natural shape of the nasal palatine floor : thus the fluid of the nose will be carried directly backward into the fauces. A piece of prepared elastic gum is next attached to the posterior part of the plate, where the natural soft palate commences, extending downward on each side as low as the remaining part of the uvula, and grooved at its lateral edges to receive the fissured portions of the velum. A movable velum is placed in the posterior centre of the elastic gum. That these may partake of the natural movements of the parts during deglutition, a spring is affixed behind them, one end of which is fastened to the posterior and anterior surfaces of the principal plate, and the other end rests gently against the posterior face of the india-rubber; this keeps it always in close apposition with the edges of the fissure during deglutition.

" It is requisite here to mention that the elastic gum should be placed in a gold frame, and not merely fastened to the posterior part of the plate, as it would shrink up by remaining in the mouth. This frame should pass round its edges only, leaving the centre open. The anterior lateral edges should be made to come considerably over the sides of the fissure, which will prevent their slipping behind it during their altered positions; the whole apparatus being held up by elastic gold springs round the teeth on each side."

1845.—Mr. Stearn, a surgeon of London, in this year communicated four articles to the *Lancet* on congenital deficiency of the palate, when he gave a description of an instrument which he had contrived for the treatment of these cases ; it was in some respects like the obturators of De la Barre and Snell, though more difficult to construct than either of them.

From a description of one constructed by the Drs. Tucker, of Boston, drawings of which we have reproduced from the eighth edition of "Harris's Dental Surgery," we are enabled to bring before our readers an accurate idea of the nature of this very interesting instrument.

It consisted of a gold plate fitted across the hard palate, having attached to it, by means of two spiral springs, an artificial velum of elastic rubber, consisting of a body, wings, and grooved edges to receive the margins of the cleft.

" Fig. 6 shows the lower surface of the palate-plate and anterior surface of the velum ; *a*, the palatine plate ; *b*, the flat spiral springs, extending from the posterior margin of the plate to the upper part of the velum ; *c c*, wings of the velum ; *d d*, the flange ; *e*, the central portion.

" Fig. 7 shows the upper surface of the palate-plate, and the posterior surface of the velum and spiral springs; *a*, the palate-plate; *b*, the spiral springs; *c c*, wings of the velum closed; *d d*, the

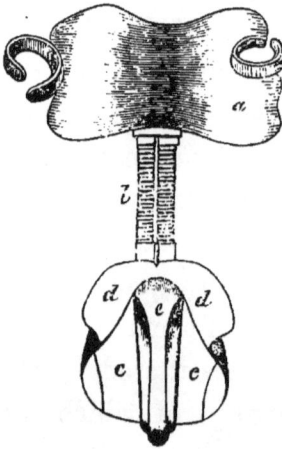 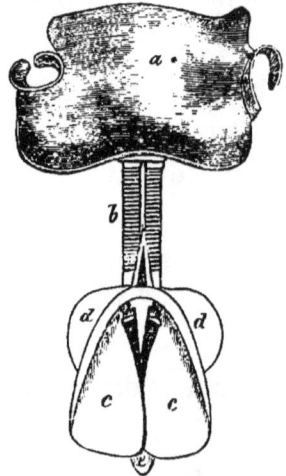

FIG. 6.   FIG. 7.

flange, as seen above the wings; and *e*, the central portion below the wings, and intended to represent the uvula.

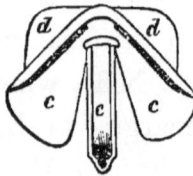

FIG. 8.

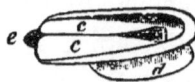

FIG. 9.

"Fig. 8 represents the velum with its wings separate from the plate, showing the central portion, before being attached to the hook, at the lower extremity of the flattened spiral springs.

"In fig. 9 is represented a side view of the velum, showing the groove between the flange and the wings, for the reception of the fleshy sides of the fissure."

1857.—Mr. Sercombe, who had for some time paid a great deal of attention to the treatment of cleft palate, in this year gave a description of the instrument he uses in remedying this defect, in a paper which he read before the Odontological Society, entitled "Cleft Palate, its Surgical and Mechanical Treatment."

From that paper we extract the following account of the instrument, with a drawing of one that had been successfully applied to a case, and worn for two years.

Mr. Sercombe says,—"My velum is made of two pieces of vulcanized india-rubber, the larger piece extremely thin, the smaller piece much thicker; the shape of both is represented in

FIG. 10.

fig. 10. The dotted line shows where they are

attached by sewing to the posterior margin of the gold plate, which has a single line of holes punched in it for this purpose. The exact size of the larger piece will vary in each case. . . . This piece should also be extremely thin, to adapt itself to the ever-varying sides of the fissure; but a piece of such tenuity as to secure this vital point, weighted with mucus, would quickly droop, but for the support which is given to it by the smaller and stouter piece which lies immediately underneath it.

"These two pieces of sheet rubber sewn to the posterior margin of the gold plate—the thinner to its upper surface, and the thicker to its lower—have been found, in more than one instance, to restore to the person using them a distinct articulation."

It will be seen from the drawing (fig. 10) that this obturator is held in position by bands passing round the molar and bicuspid teeth.

1864.—In this year Dr. Norman Kingsley, of New York, brought before the Odontological Society of Great Britain a method of treatment that for its merit demands the highest praise. The instrument itself was not altogether new in form, being to some extent very similar to that which had been constructed some years before by Mr. Stearn.

The interest attaching to the paper was rather the account of the *modus operandi*, which was briefly mentioned, the two great novelties in Dr. Kingsley's treatment consisting in taking an impression of the parts in plaster of Paris instead of wax, and preparing the elastic rubber vela in metallic moulds, rendering duplication of them a very easy matter.

The instrument, and a description of its various parts, are shown below.

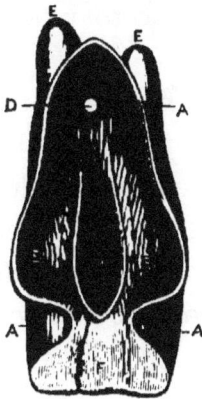 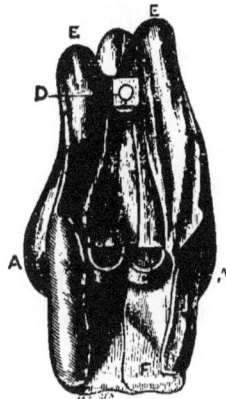

Fɪɢ. 11.          Fɪɢ. 12.

Fig. 11 represents the lower, and fig. 12 the upper part of an artificial velum. A indicates the groove in which the sides of the cleft repose; F, the posterior end, which may come in contact with the wall of the pharynx. The surface, B, lies next the tongue. G, springs of the same material, which assist it to keep its form and place. The points E rest on the top of the bone at the apex of the fissure. D, the hole through which the attachment is made, to keep it from running back.

1865.—An instrument was brought before the above-mentioned society,* made on Dr. Kingsley's principle, but much simplified in construction.

1867.—Mr. George Parkinson, in a communication to the *Lancet*, makes the following remarks on his method of treating cleft of the hard and soft palate.

(We have reproduced the drawings by which his article was illustrated, in order to render it the more interesting.)

"In a case of congenital fissure of the palate extending through the hard tissues and alveolar ridge, after having taken a correct model of the parts in wax or plaster of Paris, I commence by fitting a thin plate of gold over the vault of the palate, as far back as the posterior margin of the

FIG. 13.

Palatine surface.          Nasal surface.

---

* In a paper on the "Treatment of Congenital Cleft Palate," by Mr. Robert Ramsay.

palate bone would have extended had the bony
arch been perfect. To the posterior margin of
this plate, by means of a hinge, is attached a
velum, constructed of hard, well-polished vulcanized india-rubber formed in such a manner as
to fit the palatine surface of the remnants of the
soft palate, and allow them to glide over it in the
act of deglutition. To keep the velum in its
place, one end of a delicate gold spiral spring is
made fast to it, the other end being fixed on the
nasal surface of the gold plate representing the
hard palate. This spring must be so adjusted as
just to keep the india-rubber velum in contact
with the soft parts, and allow the portions of
uvula on either side to approximate in the act of
deglutition."

Fig. 14.

Fig. 14 represents an obturator made by Dr.
Suersen, of Hamburg, entirely of hard rubber.

D 2

A gold medal was presented to this gentleman, on account of his invention, by the Central Association of German Dentists.

It is impossible to mention all those gentlemen who, of late years especially, have treated by one means or another the defects of the organs of speech and deglutition. We can only name some of them; and trust, at a future time, to have the means of presenting our readers with an account of the special modes of treatment adopted by Dr. Bogue, of New York; MM. Préterre and Rottenstein, of Paris; and Messrs. Hulme, Vasey, Williams, and others, of London.

We have endeavoured, briefly it is true, to trace, from the first accounts given, the successive stages by which we have arrived at the present mode of treatment, showing the development of the principle that the obturator should not simply fill up the gap in a cleft palate, but be so constructed as to work on physiological principles with the natural movements of the sides of the cleft.

In 1844 Sir William Fergusson demonstrated the precise action of the muscles of the split palate; and in 1845 Mr. Stearn gave to the profession an account of an instrument which, from the movements it was capable of, we are led to conclude was constructed with a view to utilize the peculiar muscular action which the year before had been shown to exist by the first-mentioned gentleman.

This may have been simply accidental, but it is worthy of note.

In Dr. Kingsley's appliance the matter was more fully developed ; but this instrument, like Stearn's, had the fault of being too complicated for general use. We now come to the consideration of our own principle of treatment. In the main it is based on the inventions of Dr. Kingsley, though considerably modified, as will be seen in the next chapter, on "The Mode of Preparing an Artificial Velum." It is impossible for us to give one form of instrument in particular, and say that is the special pattern that we use. We endeavour in every instance to produce an obturator that will best meet the necessities of the case, not confining ourselves to one set rule, always bearing in mind, however, the important point of supplying the congenital cleft with an instrument that shall depend for its support upon the overlaps to the margins of the cleft, and not upon the teeth, having in recollection the injury that we have seen follow the attachment of any bands or wire to those very important organs of speech and mastication. This we consider a point of the greatest consequence, and one which cannot receive too much care and attention in the treatment of cleft palate by any form of obturator.

We have recently, and with most satisfactory results, attempted the imitation in the elastic

velum of all the parts that nature has left undeveloped, and the following woodcut (fig. 15)

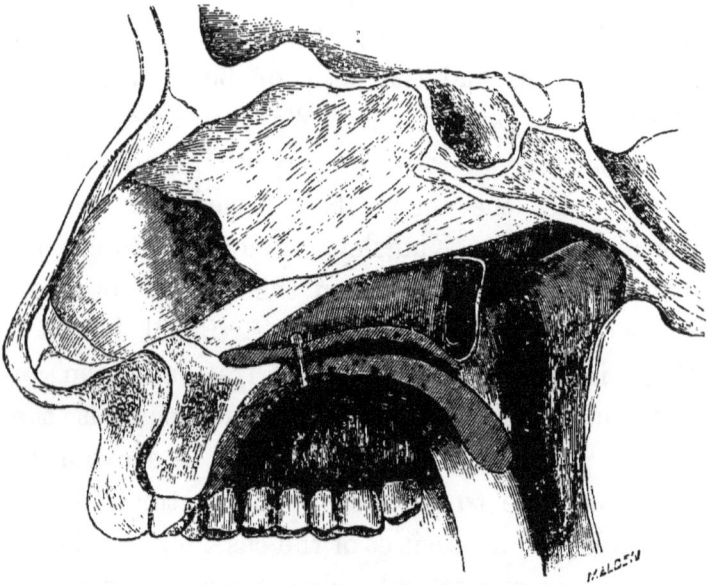

Fig. 15.

shows in section a case which is described at
page 75, in which will be seen the nasal septum,
posterior opening to the nares, with the velum
and uvula reproduced in this manner.

In the seventh chapter, containing an account
of the treatment of six cases, are given the
further variety of forms that we use under different
circumstances.

# CHAPTER V.

IT will be readily understood that in an appliance such as we have described, successful results in a great measure depend on the accuracy of the impression from which the model is made. We therefore crave the patience of our reader if we bestow what may seem at first sight an unnecessary amount of description on this part of the operation.

The materials generally used for taking impressions of the mouth are wax or some other plastic preparation, such as gutta-percha or Stent's composition ; but we think it will be admitted that these substances are by no means satisfactory, especially in taking impressions of parts that are so easily displaced as the soft palate, for none of them can be used, under the most favourable circumstances, without applying pressure sufficient to render the impression and model incorrect.

It being, then, necessary to introduce some

preparation into the mouth in such a state that it will not move the most delicate fold of mucous membrane, while in a short time it shall become so hard as to admit of removal without any alteration of form, we invariably use plaster of Paris, and so satisfied are we with the results obtained, that for even small cases of artificial teeth in the upper jaw we prefer it very much to wax or Stent's. Still, for the lower jaw, having on many occasions carefully tested it, we cannot recommend its use. In most cases the soft palate will be found too sensitive to admit of a full impression being taken at once, or even of the holding of the impression plate in position sufficiently long to admit of a model being taken. Two courses are open to the operator to overcome this difficulty: one is, to take an impression first of only the front of the mouth and cleft, and then on successive occasions gradually extend it backwards, till at last you are enabled to get a good impression of the whole of the parts, extending outwards to the alveolar ridge, upwards to the remains of the vomer, and backwards to the posterior wall of the pharynx and pillars of the fauces. Another method is to paint the parts with a solution of bromide of ammonium or tannic acid, applied with a camel's-hair brush— the brush acting almost as beneficially as the preparation used.

One or other of these two plans must be

adopted before any hope can be entertained of getting a good impression. When the parts are rendered sufficiently insensible to the presence of a foreign body, an impression-tray must be carefully prepared, so as to fit in front closely to the teeth, and at the back part leave a space about the eighth of an inch in extent from its surface to the corresponding surface of the soft palate. This does away with the necessity of an excess of plaster, and the consequent risk of any portion falling into the throat or upon the base of the tongue, and thus produce such irritation that the utmost self-control on the part of the patient will scarcely be able to overcome. The plate being in this way prepared for use, the next step is the mixing of the plaster; and here several considerations must be taken into account—(1) the dryness of the plaster, (2) its strength, and (3) the time it takes to set, which will depend partly on its freshness, and partly on the temperature of the atmosphere, as well as the water with which it is mixed.*

The best plan is to have the water with just the chill off, and then add salt in the proportion of as much as will lay upon a sixpence to half a pint.

* In a paper read before the Odontological Society, May 1st, 1865, an impression-cup made expressly of hard vulcanite was recommended; we have, however, found one modelled with gutta-percha, on an ordinary metal tray, to be quite as satisfactory and much less trouble to prepare.

If you wish the plaster to set quicker than under these circumstances it would do, add to it before mixing a small portion of rouge. This will make it set so quickly, and so strongly, that increased care and watchfulness will be required with regard to the proper time for removal from the mouth. Everything being now ready, the plaster is mixed in the ordinary manner, care of course being taken to break up all lumps in it during mixing; a sufficient quantity is then placed in or upon the impression-plate, and the whole steadily introduced into the mouth and held firmly in its place, the precaution being adopted at the moment of putting the plate in position to incline the patient's head forward, so as not only to get a good overlap above the anterior margin of the cleft, but also to lessen the liability of any plaster running down backwards and causing retching.

Now is the time to test the patient's confidence in the operator. If there is any evidence of restlessness or nervousness divert the attention by some remark, or by examining the plaster remaining in the bowl in order to ascertain the precise moment for removing the impression from the mouth,—by these or similar means to make the time (which should only occupy about a minute and a half) appear less, and save any disagreeable consequences either to yourself or the subject of your operations. To those inexperienced in these matters all this instruction may appear superfluous, but its neglect will assuredly upon many occasions lead to a decidedly "embarrassing situation."

When the remains of the unused plaster in the bowl will break asunder and leave a clean, sharp fracture, then it is time to remove the impression from the mouth. If at the first it cannot be disengaged easily, then at once and without any hesitation use sufficient force to detach it, bearing in mind that at such a time every second's delay increases the difficulty. Under ordinary circumstances it will break away in the line of the cleft. This need occasion no alarm: only desire your patient to sit perfectly still and keep the mouth well open; you can then without any anxiety or hurry push the part which remains above the margin of the palate carefully backwards to the widest part of the opening, and, firmly seizing it with a pair of long tweezers, withdraw it.

The fractured parts, when put carefully together, will be found quite as efficient for use as if no breakage had taken place, especially if, instead of using resin and wax cement, they are united with liquid silex, as recommended in the *British Journal of Dental Science* for June, 1868,* by which means any increase of bulk is avoided.

The impression, being thus perfect, must be carefully washed over with a solution of soap

* " Liquid Silex." By James Oakley Coles.

(brown Windsor is the best for the purpose), and the model made in three portions, as shown in the accompanying engraving (fig. 16). We now

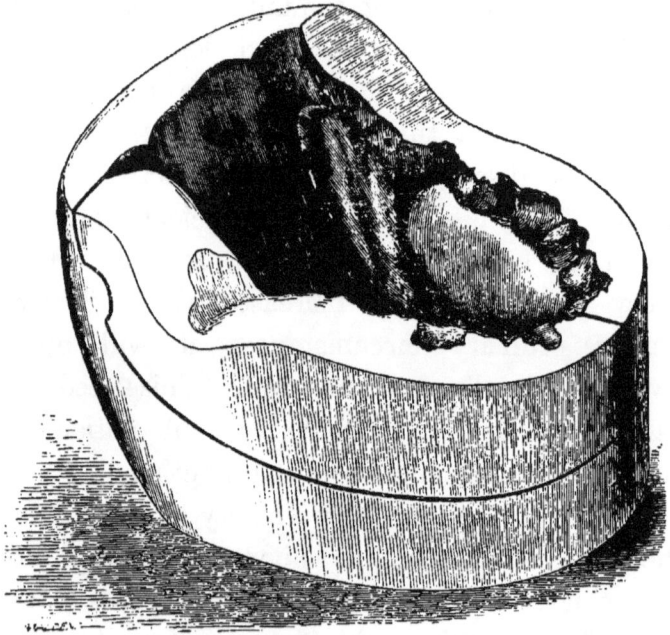

Fɪɢ. 16.

return to the more commonplace operations of the work-room, and further minute particulars would only become tedious and unnecessary.

The model being ready for use, the artificial velum must be set up in gutta-percha, having the precise shape which it will possess in its finished form. Here instruction on our part is useless, as the formation of the palate-piece wil' depend entirely on the characteristics of the case and the ingenuity of the operator. The gutta

percha should be of the best description, and the
model prepared with soapstone, to prevent any
adhesion to its surface. When this is worked up
to a satisfactory state, the casting of the plaster
moulds can be proceeded with. For an ordinary
case the best form is that shown in the engraving (fig. 17, page 62). These, however, admit of
very many modifications, according to the shape of
the velum, in preparation. The plaster castings,
when complete, must be duplicated in type metal,
the best metal obtainable and the finest casting-
sand only being used. Great care must be taken
here, as any imperfection in the metallic moulds
will be communicated to the surface of the rubber
during vulcanizing, and can only be remedied by
clipping and paring, which gives a very unsightly
appearance to the finished work. When the
castings are complete, they should fit together
accurately ; if they do not, there is no alternative
but to commence *de novo* till you arrive at a
satisfactory result.

The accompanying engraving (fig. 17) shows the
castings separated, also the metallic pin fixed in
the base for producing the hole in the velum by
which it is attached to the hard rubber front
piece. Any error with this will be found to
upset the entire arrangement. The greatest care
must therefore be used in getting it into a good
position, according to the shape of the cleft and
mouth. The moulds having been well soaped to

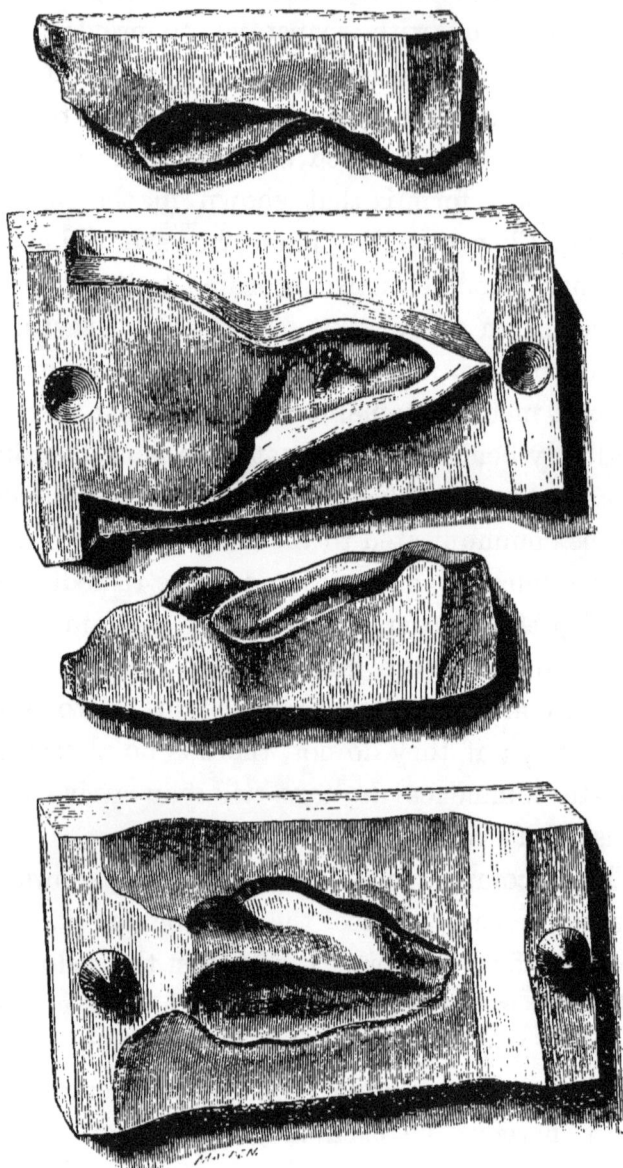

FIG. 17.

prevent adhesion, the next step is to pack them with elastic rubber. This is very easily accomplished : the two side-pieces, being adjusted to the base, are kept firmly in position by an iron clamp, and the rubber packed in from above. When there appears sufficient, the top is put on, and the whole screwed tightly together, being put on a hot plate for a few minutes to soften the rubber. The casts are then taken apart, any excess removed, or any deficiency filled up. They are again screwed up and fitted in an iron framework, with wedges to secure them, and put into the vulcanizer. In reference to the rubber to be used, there can be no question that that which is prepared by Messrs. Ash & Sons is by far the best, both as regards quality of materials and wear.

If this description of rubber be used, the time for vulcanizing is six hours ; that is to say—

> 2 hours at 240°.
>
> 2 hours at 250°.
>
> 2 hours at 260°.

This will produce an artificial velum of the greatest elasticity and power of resistance to the acids of the mouth. It has occasionally been a subject of inquiry as to the description of vulcanizer we use ; we have therefore obtained a drawing of one from the maker, Rutterford, of Poland Street, with a description of its different novelties. We use the largest size made, and place the mould for vulcanizing as near the centre as possible. Having tried several descriptions of boilers, we find this the most satisfactory (fig. 18).

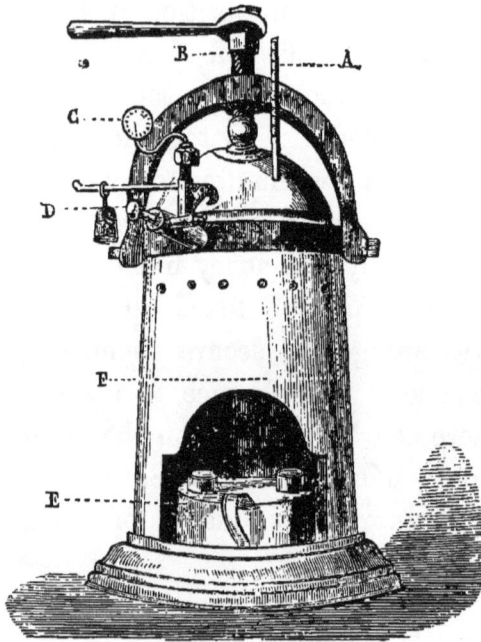

FIG. 18.

The boiler is of copper a quarter of an inch thick, and five inches in diameter. It is made for one, two, three, and four flasks, with the ordinary rings. Each boiler is properly tested and *stamped.* No washer, or india-rubber ring required. One screw only (B) to tighten the lid. D is a let-off tap for steam. The safety-valve is thereby preserved, and is not liable to get out of order. The lid contains a thick lead collar, which does not require renewing. It can be used with a steam gauge (C) if preferred, or with the ordinary thermometer (A), or both. It is equally adapted for alcohol or gas.

The adjustment of a front piece to keep the velum in the cleft will depend on the state of the teeth. If they are all perfect, a simple suction-plate, as shown in page 75, fig. 23, is all that is necessary. If any teeth should be wanting, artificial ones to supply their place should be mounted on the front piece, as in an ordinary set of teeth ; and if there be any deformity of the hard palate, as in most cases there is when associated with hare-lip, it will have to be restored and made as symmetrical as possible by additions to the hard rubber. When, however, the anterior portion of the mouth is perfect, the palate should be made as thin as possible, and not extend further back than the second bicuspids.

# CHAPTER VI.

ON THE INTRODUCTION OF THE INSTRUMENT INTO THE MOUTH. SUBSEQUENT TUITION. VALUE OF SINGING IN FACILITATING THE PROPER USE OF THE PARTS. IMPEDIMENTS TO PRODUCTION OF PERFECT VOICE.

THE artificial palate and velum being completed, the next step is its introduction into the cleft. This is sometimes the source of a little difficulty in very nervous subjects, as the presence of the foreign body (though the mouth will by this time have become less sensitive) cannot be borne under all circumstances with patience.

It is well to try in the elastic velum alone in the first instance, having it attached to a long piece of stout platina wire, one end being fitted into the pinhole previously mentioned. This will allow of its being passed well down at the back of the mouth, in order to get the wings or flaps into their proper relative position without much strain on the soft parts, while, on account of the length of the wire, the operator is able to see well what he is about. The velum can then be drawn forward into its proper place, and held there firmly for a minute, or longer, if the patient can bear it.

In most cases this produces no discomfort, while in others there is a feeling of suffocation, in consequence of the greater separation of the mouth from the nose during breathing. If excessive uneasiness be felt at any particular part, the velum must be carefully trimmed away till it becomes easy; and should the portion removed be of any great extent, new castings must be made for the metallic moulds, and another elastic rubber piece vulcanized, in order that there may be no permanent roughness to produce irritation of the mucous membrane. The fit of this portion of the instrument being satisfactory, the hard rubber front piece should be tried in, and adjusted with sufficient nicety for the patient to be able to remove and replace it at pleasure, no metallic bands or wires being used to give it greater firmness, the fact being borne in mind that the use of the front piece is not to support the velum, but to keep it in such a position that it will support itself by means of the overlap above the margins of the front or side of the cleft. When the fit of each part is considered satisfactory, they can be put together, and introduced in the complete form.

The first discomfort of wearing an instrument in the mouth having been overcome, the question arises as to whether the patients should be put under a systematic course of instruction in regard to the proper use of the tongue and soft palate, or whether they should be allowed to follow their own will and pleasure in the matter.

After some considerable thought and experience in connection with this matter, we find it impossible to lay down any fixed rule upon the subject; there are such varieties of temperament and various degrees of intelligence in the patients presented for treatment with this deformity that to attempt to be specific in advice would only lead, if followed, to annoyance and perplexity. We shall therefore only put forward such general directions as in practice we have already found beneficial, according to the different classes of subjects presented to our notice.

With the very nervous and timid patients we always recommend reading aloud, in the first instance in private, and then before a friend who shall have sufficient discretion to give such an amount of instruction as shall guide the pupil without causing any agitation in fruitless endeavours to pronounce some difficult letter or sentence.

By this means confidence is gained, and gradually the muscles of the throat and tongue will be got under control, so that not only the power to use these members properly will be acquired, but at the same time there will be the capability of preventing a return to those unnatural movements which the absence of some portion of the palate may have caused.

For those of a hopeful and vigorous turn of mind, a teacher of elocution will be of great service, while others with a quick ear for sound, and impelled by a strong feeling of pride, will make as much, and in some cases more progress, when left to themselves.

There can be no question that, under any circumstances, it is of immense advantage if you can get your patient to sing in a good loud voice with an accompaniment of some musical instrument for a quarter of an hour regularly every day. The attention is thus diverted, and all the organs of voice and articulation brought into more vigorous action than in ordinary speech, while the fact of the vocal sounds predominating over the secondary or articulate sounds will encourage the patient to persevere, through the defect not being so apparent, while at the same time the tongue and soft palate will almost instinctively assume their normal movements. That this is the case in stutterers and stammerers is a well-known fact, and our experience leads us to believe that similar results (apart from the physical deformity) follow in cleft-palate cases.

There is one very perverse habit that the tongue acquires in some instances, of applying itself constantly to the back of the lower incisor teeth, while at the same time its middle part and base are elevated in an endeavour to close the cleft in the soft palate, and thus produce a more distinct utterance. In several cases the continual pressure at the back of the incisors has caused them to separate and protrude in a very unsightly manner. This state of things has been generally accompanied by single or double hare-lip, as well as the split palate. Mr. Skey, in his second edition on "Operative Surgery," page 544, recommends a pebble being kept in the mouth, or a glass bead tied at the posterior surface of the lower front teeth, as a remedy. This is not so common in patients having simply cleft of the soft palate as when it extends into the hard portion, and there produces deformity, in so far as our own experience has shown; and after treatment by mechanical means, the presence of a foreign body in the roof of the mouth seems to have a beneficial result in regulating this unnatural movement, and causing the tongue to be placed more frequently at the back of the upper incisor teeth, and so assist the perfect articulation of the letters T, S, and others in which the sound is produced by the combined action of these two parts. Where the operation for hare-lip has not been very carefully performed, there is often a triangular space left open on the lips being approximated; just about the spot where union has taken place, as in fig. 19. This is sometimes a source of great trouble to the patient when he tries to produce the labial sounds, as it is next to impossible to get a complete closure so as to give a

clear articulation of such letters as B and P. A
consideration of this point by the surgeon would

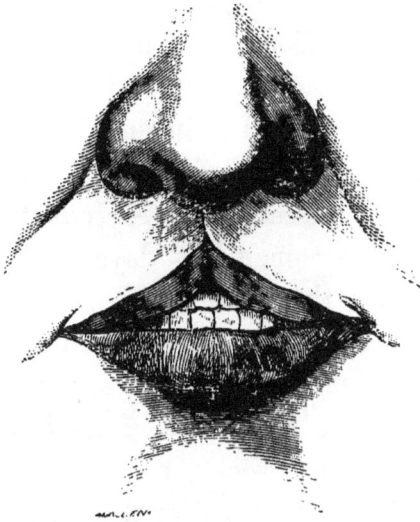

FIG. 19.

no doubt lead to some arrangement of the parts
by which this occurrence could be avoided. It
would be encroaching upon our readers' time to
enlarge upon this part of the treatment of cleft-
palate cases. The necessities and conditions of
each case will always be sufficient guide as to
whether a professed elocutionist, a friend, or
private and solitary effort is the best course to
adopt in order to produce clear and intelligible
speech.

While we, with so many others, confidently
assert the attainment of such perfection in speech
as shall allow the patient to pass through life

without discomfort to himself or attracting attention, by reason of any peculiarity, from others,
still it must not be forgotten there are instances
of such weak intellects and highly nervous tem-
peraments that no treatment in the world can
be reasonably expected to be successful.

Where the deformity of the mouth is the result
of accident or disease, the reproduction of the
lost part will within a few days, and sometimes
at once, restore the voice to its natural tone and
clearness of expression ; in congenital cases the
time which must elapse before any result can be
realized from the treatment varies from six days
to sometimes many weeks, or even a year. The
latter are, however, exceptional instances.

# CHAPTER VII.

FIRST CLASS OF CASES—CONGENITAL.

CASE 1.—Miss W——; æt. 17.—Cleft of hard
and soft palate measuring one inch and a quarter

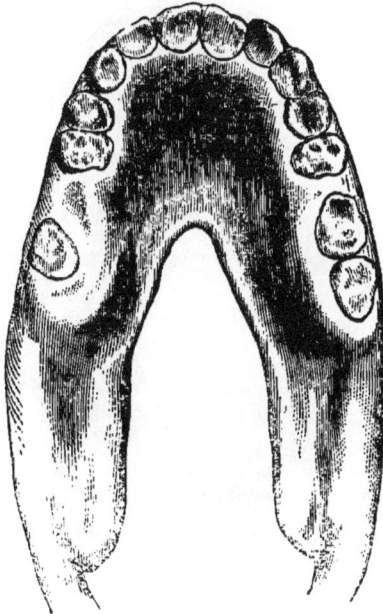

FIG. 20.

across at the widest part. This excessive width
from point to point of the uvula was mainly
owing to a very heavy gold plate, and large
hollow ball attached, also of gold, having been
worn for some time as an obturator. The teeth
were very much injured by the clasps that had
been necessary to support it. When first brought
under our notice for treatment, it presented the
appearance shown in the accompanying drawing,
the possibility of an operation being entirely pre-
cluded by the extreme size of the opening.
The model having been obtained in the usual
way with plaster of Paris, an instrument was
prepared for wear of the form shown in the an-

FIG. 21.

nexed woodcut, in this way entirely filling up the cleft and restoring the mouth to its natural condition. As this case was treated some two years and a half ago, we had not then introduced the special form of velum which we show on page 81.

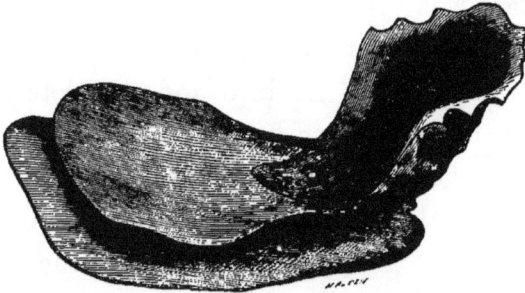

FIG. 22.

The two parts of the velum united and ready for fitting into the mouth.

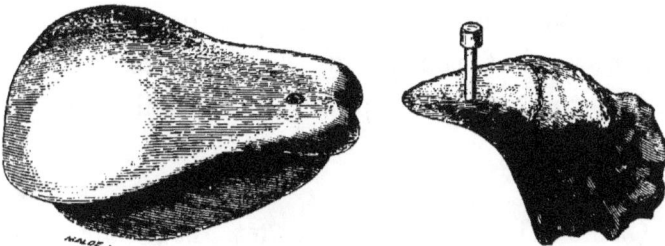

FIG. 23.

The two parts separated, showing the means by which they are held together.

CASE II.—Miss J——; æt. 16; fair complexion; very nervous temperament; small cleft in the soft palate only.

This would have succeeded admirably in the

hands of a surgeon, if the young lady would have submitted to the operation—that is, so far as the closing of the cleft; but there was also a portion of the membranes at the posterior part of the nares undeveloped, so that, although the cleft of the palate was very small, the voice was nearly as bad as in the first case mentioned. An elastic velum was made to fit into and completely close the cleft, being kept in by a full-size palate-piece. The improvement in the sound of the voice was immediate; the speech became more distinct and intelligible in a very short time, the results affording much satisfaction both to patient and operator.

Here, although the cleft was so small, still, the posterior nares being defective, surgical treatment would have accomplished only a portion of that restoration of the parts that was essential in order to give the power of intelligible expression.

CASE III.—Mr. D——; æt. 19; fair complexion; nervous temperament.—Cleft of hard and soft palate, extending also through the dental arch, a single hare-lip having been treated in a very satisfactory manner soon after birth.

The left side of the maxillary portion of the fissure at its border was continuous with the vomer, thus giving no overlap except on one side and at the apex of the cleft. The following

woodcut taken from a model of the mouth shows
the appearance before treatment :—

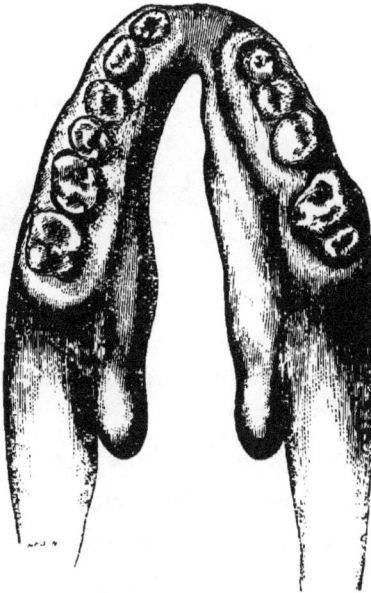

Fig. 24.

In these cases the incisors are generally but
very imperfectly developed, we therefore removed
the centrals and considerably improved the
mouth for the purposes of speech and mastication by fitting artificial teeth in their place, on
the front piece of the velum. The artificial palate
was then made, having the form shown in fig. 25.
This was put in November, 1865, and has been
worn with great comfort ever since. In less than
a twelvemonth the articulation of every word and
sound was perfect; and those most difficult
letters, K and R, were pronounced with the

greatest ease and precision. The only thing
that could not be overcome was the peculiar

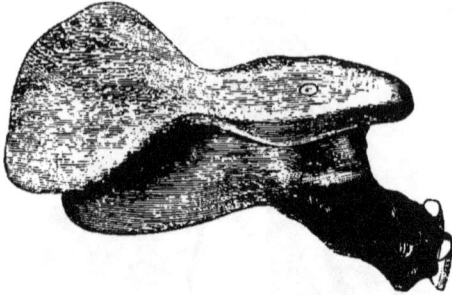

FIG. 25.

The velum and hard rubber front piece ready for fitting into
the mouth, showing all the overlap obtained, being on one
side only and at the front.

nasal tone, very slight however, that results
always from the malformation that occurs
in the nose after the operation for hare-lip.
Before the patient was put under treatment, his
speech was quite unintelligible, even to many of
his friends; yet after the mouth was restored to
as symmetrical a form as under the circumstances
was possible (see fig. 26), this gentleman by
his own perseverance, unassisted by any one's
tuition, acquired such clearness of utterance as to
elicit expressions of the greatest astonishment at
both the Medical and Chirurgical and Odontological
Societies, before whose members he very kindly
went through the alphabet and several difficult
sentences. Within a short time of this he was
elected an officer in a volunteer rifle corps, as well
as a volunteer fire brigade, thus affording the
best evidence of the ability he possessed to make

himself readily and perfectly understood by any one with whom he might come into contact.

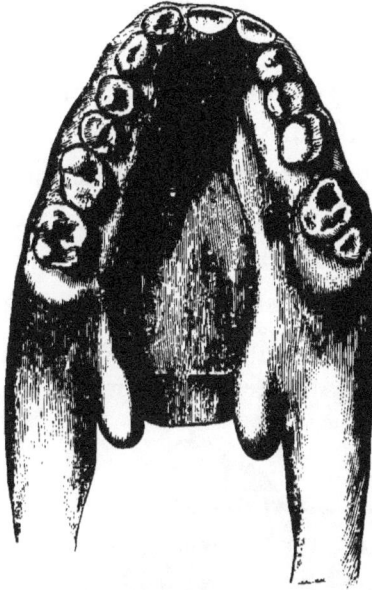

FIG. 26.

The following drawing shows the two parts of the instrument separated, and also gives indication of the manner in which the hard and soft palates were restored to a proper form :—

FIG. 27.

CASE IV.—Miss F——; æt. 17.—This case resembled all the others as regards temperament

and complexion, though there was not much
sensitiveness as regards the deformity, and unfortunately no ear for musical sounds, though the
young lady played several instruments with ordinary accuracy and ability. There was also slight
deafness, probably arising from inflammation
of the mucous membrane around the Eustachian
tubes, the inflammation having arisen from the
great exposure of the parts to every change of
temperature in consequence of the opening in the
palate. The mouth, when presented for treatment, had the appearance shown in fig. 28.

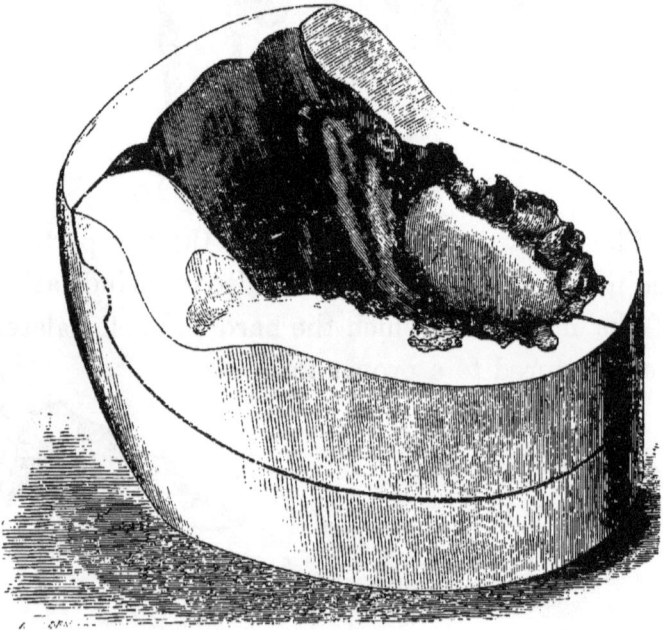

Fig. 28.

A velum was made which restored the uvula
in the lower flap, and in upper flap reproduced the

septum of the interior nares where it was absent, also the posterior nares with its two openings.

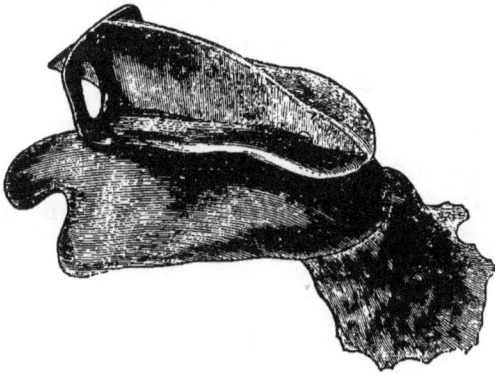

FIG. 29.

The artificial velum and front piece attached by means of the platina pin.

By these means the mouth, nose, and upper part of the pharynx were restored to their natural

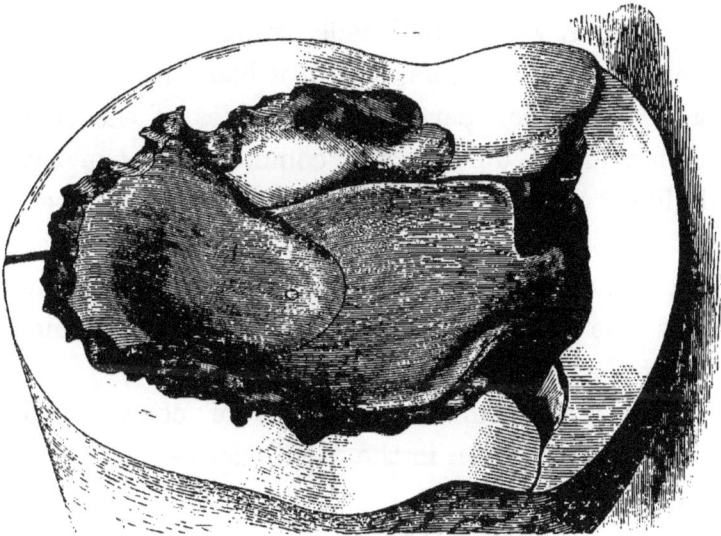

FIG. 30.

The mouth, as artificially restored.

condition, and much satisfaction was afforded by the improvement in a very short time, not only in the facility with which the patient could make herself understood, but also in the tone of the voice, which was unquestionably owing to the alterations that had been produced in the form of the superior part of the pharynx.

Up to the present time the progress in this case has been steady and satisfactory.

CASE V.—Mr. R——; æt. 22; cleft of the soft palate, extending just beyond the posterior margin of the hard palate; treated May, 1864.— The voice in this case was very bad. The young gentleman was most anxious to have something done, as a public appointment was being kept for him, provided his speech could be rendered intelligible. A velum was made fitting into the cleft, but, unlike the others shown, it had only one flap at the posterior part, the two sides of the cleft embracing it only to the commencement of the bulbous portion of the bifurcated uvula. Though this extremely simple form of instrument was used, the result was such that in two months he was able to enter upon the duties of the appointment that had been held open for him.

The elastic rubber piece was of course held forward in position in the usual way:—

CASE VI.—Mr. B——; æt. 15; fair complexion, nervous temperament ; cleft of hard and soft palate, the division extending through the maxillary bones, with an opening from the nostril to the mouth between the lip and the labial surface of the alveolus, the operation for hare-lip having been performed during infancy. —This young gentleman's speech was so exceedingly defective that he could only commmunicate with strangers by means of writing, while those who were well acquainted with him had to pay the greatest attention to understand what he attempted to say.

The mouth was treated (being very similar in form) in nearly the same way as Case No. 3, with the addition of a process (if we may apply an anatomical term to such a purpose) to run forward into the left nostril to prevent the passage of air between the front of the mouth and nose in the utterance of the explosive sounds, when the lips are firmly compressed, and the air for a moment requires confinement in the cavity of the mouth. This case proceeded very satisfactorily, the voice and speech improving in a very short time, the health also being benefitted from the greater care with which the food was prepared for swallowing.

# CHAPTER VIII.

SIX CASES OF THE SECOND CLASS OF DEFORMITIES—
THOSE ARISING FROM ACCIDENTAL CAUSES—
TREATED BY MECHANICAL MEANS.

CASE I.—Sophia S—— ; æt. 32.—Applied at the Hospital for Diseases of the Throat for treatment of severe ulceration and loss of parts at the back of the mouth. Nearly the whole of the velum palati had disappeared, the anterior and posterior pillars of the fauces were likewise destroyed, so that the roof of the mouth presented the appearance of continuance backwards to the posterior wall of the pharynx. In the position that would be occupied by the uvula and central portion of the soft palate, when elevated for dividing the mouth from the nose, there was a large opening of an oval form, about one and a quarter of an inch in extent one way, and three-quarters of an inch from side to side. In swallowing, there was not the slightest movement at the back of the mouth, except in the tongue, which was the only member that could contribute any assistance to the process of conveying the food to the opening into the œsophagus. The back of the mouth was in this way kept in a very irritable condition by the continual lodgment of food in the cleft. From the state of the palate, speech was scarcely intelligible, and the life of the poor woman was in every way a matter of considerable discomfort. Owing also to the great induration of the parts on each side, where the indications of the anterior pillars of the fauces were apparent, we concluded that no power could be obtained to work an elastic velum with any service or comfort, while at the same time there was the consideration to be borne in mind that the disease was still going on, and it was desirable rather to protect the parts from the irritation resulting from food, &c., than to increase the trouble by having an artificial velum, that must necessarily produce some chafing, the mucous membrane being so exceedingly sensitive. A simple hard rubber obturator was therefore made, partially closing the aperture, and having the inner surface highly polished. This has been very satisfactory in its results.

CASE II.—William T——, engineer; æt. 37.— In this case the upper maxillary bone was destroyed on the left side from the central tooth to the second molar tooth, following the line of the intermaxillary suture, and the connection of the palate-bone with the upper maxilla. The septum of the nose was quite perfect, articulating with the maxillary bone of the opposite side.

The turbinated bones of the left side, with the walls of the antrum, were entirely destroyed up to the floor of the orbit, leaving a gap for restoration by artificial means of considerable extent. The voice was very imperfect, mastication and swallowing very difficult. The appearance of the case is shown in fig. 31.

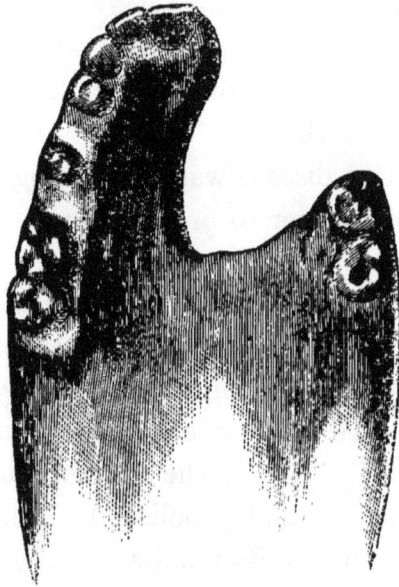

FIG. 31.

The instrument that was constructed to remedy these defects is shown complete, ready for wear, in fig. 32; also with the parts separated, showing how far the hard rubber extended, and where the elastic india-rubber was connected with it, in order that the more delicate parts might not be irritated.

The means that were adopted were not only
satisfactory, but immediate in their result—speech

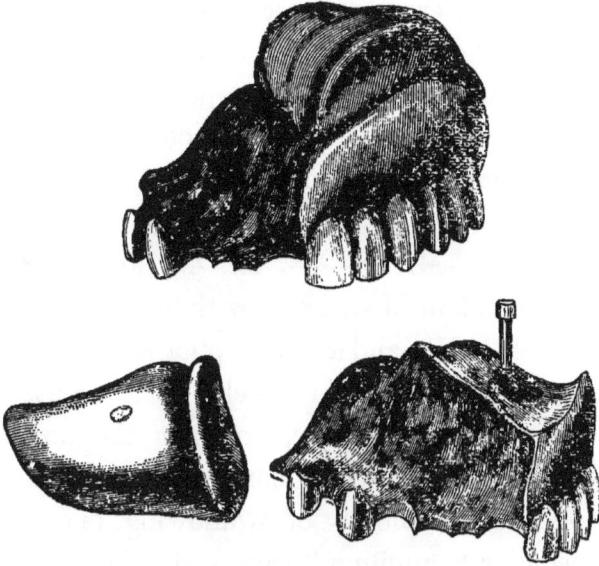

FIG. 32.

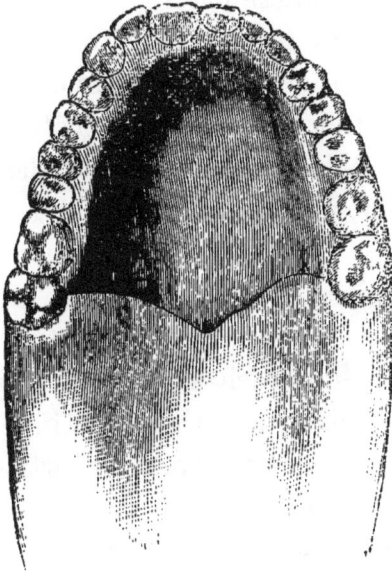

FIG. 33.

was restored at once to its normal tone and distinctness. Gargling the throat and mouth (before impossible) were now accomplished with ease, while by the restoration of the teeth to their natural state the patient's appearance was very much improved. The appearance of the mouth after treatment is shown in fig. 33.

Case III.—Mr. B——, æt. 43, applied to us for treatment in May, 1866, with loss of the whole of the soft palate and anterior pillars of the fauces. The pharynx above the level of the palate had not received any injury, so that the posterior nares were in a perfect condition. The patient complained of difficulty of swallowing, and also of the peculiar "snuffling" tone which his voice had since the loss of these parts.

The treatment in this case was comparatively simple : the usual plaster impression having been obtained, a full palate suction-plate was made, with a soft palate reproduced in elastic rubber attached to it, passing backward in the horizontal (not pendent) position, so that when at rest it was at its greatest elevation, and could be depressed only by the action of the remaining muscular fibres on each side in the region of the fauces. By this means there was in a short time acquired such control over the artificial velum palati that it answered all the purposes of the natural member which had been lost.

CASE IV.—Mr. William D——, æt. 46, came to us in December, 1865, with perforation of the soft palate of the form and position shown in fig. 34, the greater part of the uvula having by our advice been removed.

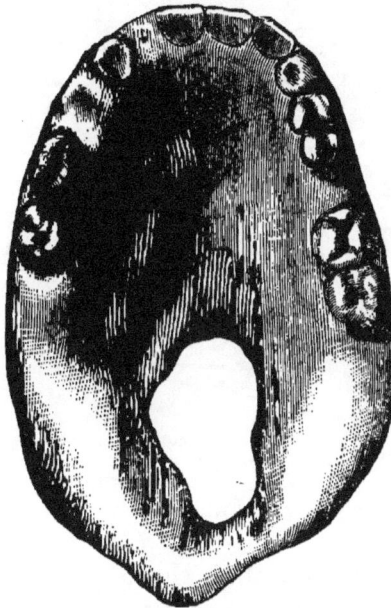

FIG. 34.

In this instance great suffering was caused by the portions of food that became lodged upon the posterior margin of the cleft, causing considerable difficulty in swallowing ; the voice was very imperfect from the rushing of air through the gap in the palate into the nose, and the throat was kept in a state of continual irritation from the great passage of cold air in the reverse direction— that is, from the nostrils into the back of the mouth. An instrument was made to close up this space in the manner shown in fig. 35.

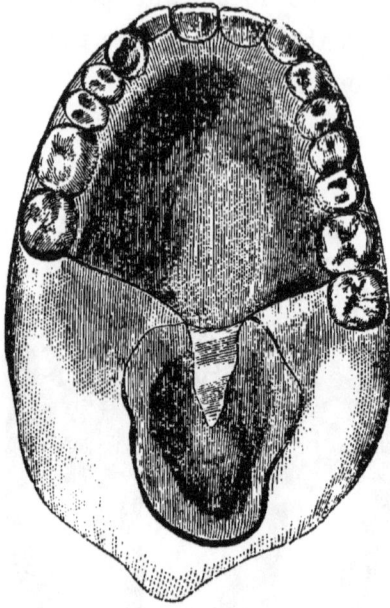

FIG. 35.

It was necessary that there should be perfect closure, and yet the power of depressing the elastic part of the obturator with the rest of the soft palate. The instrument was therefore constructed in the way shown in fig. 36. The front portion

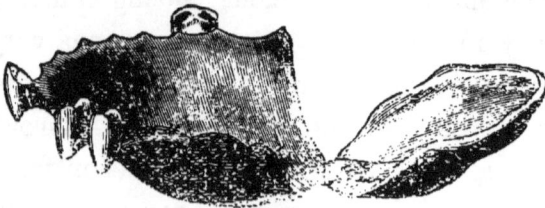

FIG. 36.

carrying the teeth that were out, of hard rubber, and a slip of gold running backwards to support the elastic rubber plug.    Fig. 37 shows the two portions of the instrument apart.

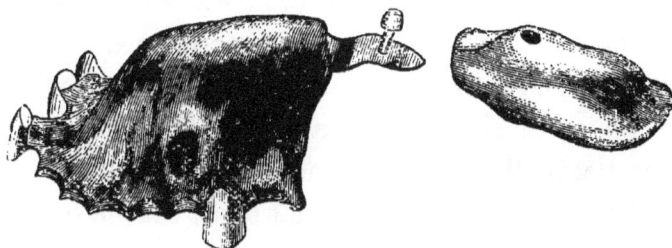

Fig. 37.

This appliance answered all the purposes that were desired and necessary, and after its introduction into the mouth the patient not only had the satisfaction of finding his speech and power of swallowing restored, but also his general health very much improved, from the absence of the irritation that his defect had previously given rise to.

Case V.—Mrs. W——, æt. 29, sent to us for treatment in September, 1867.   This case need only be alluded to very briefly, as it closely resembled the last one described (Case 4).

The only difference was that the opening in the soft palate was lower down, and to the right side, instead of being nearly in the centre of the palate.   The difference in treatment consisted in conveying a flap of elastic rubber back in place of the strip of gold mentioned in the last case, so as to give greater mobility to all the parts, and thus do away with any risk of setting up irritation again. This expedient of substituting rubber for gold on this account was the more advisable, as the uvula was only connected on one side by a mere thread of mucous membrane, while on the other side it had a tolerable attachment. The elastic flap was connected to a suction-plate in front carrying two artificial teeth that had been lost some time previously.

In this case also the results were as satisfactory, though not so rapid, as in all the others mentioned. We much wished the uvula removed, but could not prevail on the patient to consent to this, though it would have considerably improved the condition of the mouth.

CASE VI.—Mr. Thomas P—— ; æt. 34 ; perforation of the soft palate, and adhesion of its free border to the posterior wall of the pharynx, with destruction of both anterior and posterior pillars of the fauces. The position of this opening in regard to its proximity to the back of the pharynx renders it very similar to the first case mentioned (see page 84). The troubles that arose from the defect were the lodgment of food above the margins, and the regurgitation of liquids into the nose, unless great care was used in swallowing ; also great difficulty in expression, especially of certain sounds, such as capers, peas, potatoes, the letters K and G, &c.

The appearance of the mouth when presented for treatment is shown in fig. 38. There was

Fig. 38.

such rigidity in the whole of the parts that no movement could be obtained by any means in

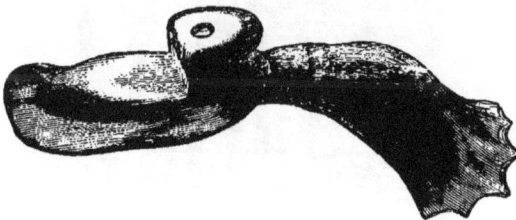

Fig. 39.

order to work an artificial palate. An arrangement was therefore made of the form shown in
fig. 39 ; the elastic portion, fitting into the
anterior third of the cleft or perforation, was
kept in place by the small suction-plate.

The flap acting as a valve gave great comfort
in mastication and swallowing, and, by being
closed from the pressure of air from the lungs,
admitted of the utterance of the explosive sounds
with great facility, while it also allowed of the
passage of air and mucus from the posterior
nares. Fig. 40 shows the mouth after the perforation has been closed by the obturator.

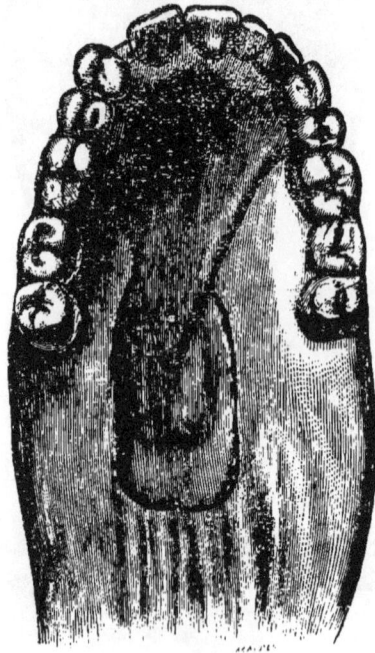

FIG. 40.

An account of many more cases of equal interest might be given, but that we think those we have described exemplify nearly every variety of ordinary occurrence. It only remains to mention those deformities that sometimes occur in the hard palate; consisting of an extreme height of the palatine arch, by which the speech is rendered very indistinct. The treatment must depend entirely upon the form the superior maxilla has assumed during the process of development, and an endeavour be made to restore the mouth, by means of an artificial plate of hard vulcanite, to as symmetrical a form as the case will admit of. This plan, though very simple, we have always found satisfactory.

WYMAN AND SONS, PRINTERS, GREAT QUEEN STREET, LONDON, W.C.

London, New Burlington Street,
October, 1868.

# MESSRS. CHURCHILL & SONS'

## Publications,

IN

# MEDICINE

### AND THE VARIOUS BRANCHES OF

# NATURAL SCIENCE.

# A CLASSIFIED INDEX

TO

# MESSRS. CHURCHILL & SONS' CATALOGUE.

a 2

TO BE COMPLETED IN TWELVE PARTS, 4TO., AT 7s. 6d. PER PART.

PARTS I. & II. NOW READY.

# A DESCRIPTIVE TREATISE

ON THE

# NERVOUS SYSTEM OF MAN,

WITH THE MANNER OF DISSECTING IT.

## By LUDOVIC HIRSCHFELD,

DOCTOR OF MEDICINE OF THE UNIVERSITIES OF PARIS AND WARSAW, PROFESSOR OF ANATOMY TO THE FACULTY OF MEDICINE OF WARSAW;

Edited in English (from the French Edition of 1866)

## By ALEXANDER MASON MACDOUGAL, F.R.C.S.,

WITH

## AN ATLAS OF ARTISTICALLY-COLOURED ILLUSTRATIONS,

Embracing the Anatomy of the entire Cerebro-Spinal and Sympathetic Nervous Centres and Distributions in their accurate relations with all the important Constituent Parts of the Human Economy, and embodied in a series of 56 Single and 9 Double Plates, comprising 197 Illustrations,

Designed from Dissections prepared by the Author, and Drawn on Stone by

## J. B. LÉVEILLÉ.

## MR. ACTON, M.R.C.S.

### I.

**A PRACTICAL TREATISE ON DISEASES OF THE URINARY** AND GENERATIVE ORGANS IN BOTH SEXES. Third Edition. 8vo. cloth, £1. 1s. With Plates, £1. 11s. 6d. The Plates alone, limp cloth, 10s. 6d.

### II.

**THE FUNCTIONS AND DISORDERS OF THE REPRODUC-** TIVE ORGANS IN CHILDHOOD, YOUTH, ADULT AGE, AND ADVANCED LIFE, considered in their Physiological, Social, and Moral Relations. Fourth Edition. 8vo. cloth, 10s. 6d.

### III.

**PROSTITUTION**: Considered in its Moral, Social, and Sanitary Bearings, with a View to its Amelioration and Regulation. 8vo. cloth, 10s. 6d.

## DR. ADAMS, A.M.

**A TREATISE ON RHEUMATIC GOUT; OR, CHRONIC** RHEUMATIC ARTHRITIS. 8vo. cloth, with a Quarto Atlas of Plates, 21s.

## MR. WILLIAM ADAMS, F.R.C.S.

### I.

**ON THE PATHOLOGY AND TREATMENT OF LATERAL** AND OTHER FORMS OF CURVATURE OF THE SPINE. With Plates. 8vo. cloth, 10s. 6d.

### II.

**CLUBFOOT**: its Causes, Pathology, and Treatment. Jacksonian Prize Essay for 1864. With 100 Engravings. 8vo. cloth, 12s.

### III.

**ON THE REPARATIVE PROCESS IN HUMAN TENDONS** AFTER SUBCUTANEOUS DIVISION FOR THE CURE OF DEFORMITIES. With Plates. 8vo. cloth, 6s.

### IV.

**SKETCH OF THE PRINCIPLES AND PRACTICE OF** SUBCUTANEOUS SURGERY. 8vo. cloth, 2s. 6d.

## DR. WILLIAM ADDISON, F.R.S.

### I.

**CELL THERAPEUTICS.** 8vo. cloth, 4s.

### II.

**ON HEALTHY AND DISEASED STRUCTURE,** AND THE TRUE PRINCIPLES OF TREATMENT FOR THE CURE OF DISEASE, ESPECIALLY CONSUMPTION AND SCROFULA, founded on MICROSCOPICAL ANALYSIS. 8vo. cloth, 12s.

## DR. ALDIS.

**AN INTRODUCTION TO HOSPITAL PRACTICE IN VARIOUS** COMPLAINTS; with Remarks on their Pathology and Treatment. 8vo. cloth, 5s. 6d.

## DR. SOMERVILLE SCOTT ALISON, M.D.EDIN., F.R.C.P.

**THE PHYSICAL EXAMINATION OF THE CHEST IN PUL-** MONARY CONSUMPTION, AND ITS INTERCURRENT DISEASES. With Engravings. 8vo. cloth, 12s.

## DR. ALTHAUS, M.D., M.R.C.P.

ON EPILEPSY, HYSTERIA, AND ATAXY. Cr. 8vo. cloth, 4s.

THE ANATOMICAL REMEMBRANCER; OR, COMPLETE
POCKET ANATOMIST. Sixth Edition, carefully Revised. 32mo. cloth, 3s. 6d.

## DR. McCALL ANDERSON, M.D.

### I.
PARASITIC AFFECTIONS OF THE SKIN. With Engravings.
8vo. cloth, 5s.        II.
ECZEMA. Second Edition. 8vo. cloth, 6s.
### III.
PSORIASIS AND LEPRA. With Chromo-lithograph. 8vo. cloth, 5s.

## DR. ANDREW ANDERSON, M.D.
TEN LECTURES INTRODUCTORY TO THE STUDY OF FEVER.
Post 8vo. cloth, 5s.

## DR. ARLIDGE.
ON THE STATE OF LUNACY AND THE LEGAL PROVISION
FOR THE INSANE; with Observations on the Construction and Organisation of
Asylums. 8vo. cloth, 7s.

## DR. ALEXANDER ARMSTRONG, R.N.
OBSERVATIONS ON NAVAL HYGIENE AND SCURVY.
More particularly as the latter appeared during a Polar Voyage. 8vo. cloth, 5s.

## MR. T. J. ASHTON.
### I.
ON THE DISEASES, INJURIES, AND MALFORMATIONS
OF THE RECTUM AND ANUS. Fourth Edition. 8vo. cloth, 8s.
### II.
PROLAPSUS, FISTULA IN ANO, AND HÆMORRHOIDAL
AFFECTIONS; their Pathology and Treatment. Second Edition. Post 8vo. cloth 2s. 6d.

## MR. THOS. J. AUSTIN, M.R.C.S.ENG.
A PRACTICAL ACCOUNT OF GENERAL PARALYSIS:
Its Mental and Physical Symptoms, Statistics, Causes, Seat, and Treatment. 8vo. cloth, 6s.

## DR. THOMAS BALLARD, M.D.
A NEW AND RATIONAL EXPLANATION OF THE DIS-
EASES PECULIAR TO INFANTS AND MOTHERS; with obvious Suggestion
for their Prevention and Cure. Post 8vo. cloth, 4s. 6d.

### DR. BARCLAY.

I.

**A MANUAL OF MEDICAL DIAGNOSIS.** Second Edition.
Foolscap 8vo. cloth, 8s. 6d.              II.

**MEDICAL ERRORS.**—Fallacies connected with the Application of the
Inductive Method of Reasoning to the Science of Medicine. Post 8vo. cloth, 5s.

III.

**GOUT AND RHEUMATISM IN RELATION TO DISEASE**
OF THE HEART. Post 8vo. cloth, 5s.

### DR. T. HERBERT BARKER, M.D., F.R.S., & MR. ERNEST EDWARDS, B.A.

**PHOTOGRAPHS OF EMINENT MEDICAL MEN**, with brief
Analytical Notices of their Works. Vol. I. (24 Portraits), 4to. cloth, 24s.

### DR. BARLOW.

**A MANUAL OF THE PRACTICE OF MEDICINE.** Second
Edition. Fcap. 8vo. cloth, 12s. 6d.

### DR. BARNES.

**THE PHYSIOLOGY AND TREATMENT OF PLACENTA**
PRÆVIA; being the Lettsomian Lectures on Midwifery for 1857. Post 8vo. cloth, 6s.

### DR. BASCOME.

**A HISTORY OF EPIDEMIC PESTILENCES, FROM THE**
EARLIEST AGES. 8vo. cloth, 8s.

### DR. BASHAM.

**ON DROPSY, AND ITS CONNECTION WITH DISEASES OF**
THE KIDNEYS, HEART, LUNGS AND LIVER. With 16 Plates. Third
Edition. 8vo. cloth, 12s. 6d.

### MR. H. F. BAXTER, M.R.C.S.L.

**ON ORGANIC POLARITY;** showing a Connexion to exist between
Organic Forces and Ordinary Polar Forces. Crown 8vo. cloth, 5s.

### MR. BATEMAN.

**MAGNACOPIA:** A Practical Library of Profitable Knowledge, commu-
nicating the general Minutiæ of Chemical and Pharmaceutic Routine, together with the
generality of Secret Forms of Preparations. Third Edition. 18mo. 6s.

### MR. LIONEL J. BEALE, M.R.C.S.

I.

**THE LAWS OF HEALTH IN THEIR RELATIONS TO MIND**
AND BODY. A Series of Letters from an Old Practitioner to a Patient. Post 8vo.
cloth, 7s. 6d.              II.

**HEALTH AND DISEASE, IN CONNECTION WITH THE**
GENERAL PRINCIPLES OF HYGIENE. Fcap. 8vo., 2s. 6d.

### DR. BEALE, F.R.S.

#### I.

URINE, URINARY DEPOSITS, AND CALCULI: and on the Treatment of Urinary Diseases. Numerous Engravings. Second Edition, much Enlarged. Post 8vo. cloth, 8s. 6d.

#### II.

THE MICROSCOPE, IN ITS APPLICATION TO PRACTICAL MEDICINE. Third Edition. With 58 Plates. 8vo. cloth, 16s.

#### III.

ILLUSTRATIONS OF THE SALTS OF URINE, URINARY DEPOSITS, and CALCULI. 37 Plates, containing upwards of 170 Figures copied from Nature, with descriptive Letterpress. 8vo. cloth, 9s. 6d.

### MR. BEASLEY.

#### I.

THE BOOK OF PRESCRIPTIONS; containing 3000 Prescriptions. Collected from the Practice of the most eminent Physicians and Surgeons, English and Foreign. Third Edition. 18mo. cloth, 6s.

#### II.

THE DRUGGIST'S GENERAL RECEIPT-BOOK: comprising a copious Veterinary Formulary and Table of Veterinary Materia Medica; Patent and Proprietary Medicines, Druggists' Nostrums, &c.; Perfumery, Skin Cosmetics, Hair Cosmetics, and Teeth Cosmetics; Beverages, Dietetic Articles, and Condiments; Trade Chemicals, Miscellaneous Preparations and Compounds used in the Arts, &c.; with useful Memoranda and Tables. Sixth Edition. 18mo. cloth, 6s.

#### III.

THE POCKET FORMULARY AND SYNOPSIS OF THE BRITISH AND FOREIGN PHARMACOPŒIAS; comprising standard and approved Formulæ for the Preparations and Compounds employed in Medical Practice. Eighth Edition, corrected and enlarged. 18mo. cloth, 6s.

### DR. HENRY BENNET.

#### I.

A PRACTICAL TREATISE ON INFLAMMATION AND OTHER DISEASES OF THE UTERUS. Fourth Edition, revised, with Additions. 8vo. cloth, 16s.

#### II.

A REVIEW OF THE PRESENT STATE (1856) OF UTERINE PATHOLOGY. 8vo. cloth, 4s.

#### III.

WINTER IN THE SOUTH OF EUROPE; OR, MENTONE, THE RIVIERA, CORSICA, SICILY, AND BIARRITZ, AS WINTER CLIMATES. Third Edition, with numerous Plates, Maps, and Wood Engravings. Post 8vo. cloth, 10s. 6d.

### PROFESSOR BENTLEY, F.L.S.

A MANUAL OF BOTANY. With nearly 1,200 Engravings on Wood. Fcap. 8vo. cloth, 12s. 6d.

### DR. BERNAYS.

NOTES FOR STUDENTS IN CHEMISTRY; being a Syllabus compiled from the Manuals of Miller, Fownes, Berzelius, Gerhardt, Gorup-Besanez, &c. Fourth Edition. Fcap. 8vo. cloth, 3s.

### MR. HENRY HEATHER BIGG.

ORTHOPRAXY: the Mechanical Treatment of Deformities, Debilities, and Deficiencies of the Human Frame. With Engravings. Post 8vo. cloth, 10s.

### DR. S. B. BIRCH, M.D., M.R.C.P.

**I.**

OXYGEN; ITS ACTION, USE, AND VALUE IN THE TREATMENT OF VARIOUS DISEASES OTHERWISE INCURABLE OR VERY INTRACTABLE. Second Edition. Post 8vo. cloth, 3s. 6d.

**II.**

CONSTIPATED BOWELS: the Various Causes and the Different Means of Cure. Third Edition. Post 8vo. cloth, 3s. 6d.

### DR. GOLDING BIRD, F.R.S.

URINARY DEPOSITS; THEIR DIAGNOSIS, PATHOLOGY, AND THERAPEUTICAL INDICATIONS. With Engravings. Fifth Edition. Edited by E. Lloyd Birkett, M.D. Post 8vo. cloth, 10s. 6d.

### MR. BISHOP, F.R.S.

**I.**

ON DEFORMITIES OF THE HUMAN BODY, their Pathology and Treatment. With Engravings on Wood. 8vo. cloth, 10s.

**II.**

ON ARTICULATE SOUNDS, AND ON THE CAUSES AND CURE OF IMPEDIMENTS OF SPEECH. 8vo. cloth, 4s.

### MR. P. HINCKES BIRD, F.R.C.S.

PRACTICAL TREATISE ON THE DISEASES OF CHILDREN AND INFANTS AT THE BREAST. Translated from the French of M. Bouchut, with Notes and Additions. 8vo. cloth. 20s.

### MR. BLAINE.

OUTLINES OF THE VETERINARY ART; OR, A TREATISE ON THE ANATOMY, PHYSIOLOGY, AND DISEASES OF THE HORSE, NEAT CATTLE, AND SHEEP. Seventh Edition. By Charles Steel, M.R.C.V.S.L. With Plates. 8vo. cloth, 18s.

### MR. BLOXAM.

CHEMISTRY, INORGANIC AND ORGANIC; with Experiments and a Comparison of Equivalent and Molecular Formulæ. With 276 Engravings on Wood. 8vo. cloth, 16s.

### DR. BOURGUIGNON.

ON THE CATTLE PLAGUE; OR, CONTAGIOUS TYPHUS IN HORNED CATTLE: its History, Origin, Description, and Treatment. Post 8vo. 5s.

### MR. JOHN E. BOWMAN, & MR. C. L. BLOXAM.

**I.**

PRACTICAL CHEMISTRY, including Analysis. With numerous Illustrations on Wood. Fifth Edition. Foolscap 8vo. cloth, 6s. 6d.

**II.**

MEDICAL CHEMISTRY; with Illustrations on Wood. Fourth Edition, carefully revised. Fcap. 8vo. cloth, 6s. 6d.

### DR. JAMES BRIGHT.

ON DISEASES OF THE HEART, LUNGS, & AIR PASSAGES; with a Review of the several Climates recommended in these Affections. Third Edition. Post 8vo. cloth, 9s.

### DR. BRINTON, F.R.S.
#### I.
THE DISEASES OF THE STOMACH, with an Introduction on its Anatomy and Physiology; being Lectures delivered at St. Thomas's Hospital. Second Edition. 8vo. cloth, 10s. 6d.
#### II.
INTESTINAL OBSTRUCTION. Edited by DR. BUZZARD. Post 8vo. cloth, 5s.

### MR. BERNARD E. BRODHURST, F.R.C.S.
#### I.
CURVATURES OF THE SPINE: their Causes, Symptoms, Pathology, and Treatment. Second Edition. Roy. 8vo. cloth, with Engravings, 7s. 6d.
#### II.
ON THE NATURE AND TREATMENT OF CLUBFOOT AND ANALOGOUS DISTORTIONS involving the TIBIO-TARSAL ARTICULATION. With Engravings on Wood. 8vo. cloth, 4s. 6d.
#### III.
PRACTICAL OBSERVATIONS ON THE DISEASES OF THE JOINTS INVOLVING ANCHYLOSIS, and on the TREATMENT for the RESTORATION of MOTION. Third Edition, much enlarged, 8vo. cloth, 4s. 6d.

### MR. BROOKE, M.A., M.B., F.R.S.
ELEMENTS OF NATURAL PHILOSOPHY. Based on the Work of the late Dr. Golding Bird. Sixth Edition. With 700 Engravings. Fcap. 8vo. cloth, 12s. 6d.

### DR. T. L. BRUNTON, B.SC., M.B.
ON DIGITALIS. With some Observations on the Urine. Fcap. 8vo. cloth, 4s. 6d.

### MR. THOMAS BRYANT, F.R.C.S.
#### I.
ON THE DISEASES AND INJURIES OF THE JOINTS. CLINICAL AND PATHOLOGICAL OBSERVATIONS. Post 8vo. cloth, 7s. 6d.
#### II.
THE SURGICAL DISEASES OF CHILDREN. The Lettsomian Lectures, delivered March, 1863. Post 8vo. cloth, 5s.
#### III.
CLINICAL SURGERY. Parts I. to VII. 8vo., 3s. 6d. each.

### DR. BUCKLE, M.D., L.R.C.P.LOND.
VITAL AND ECONOMICAL STATISTICS OF THE HOSPITALS, INFIRMARIES, &c., OF ENGLAND AND WALES. Royal 8vo. 5s.

### DR. JOHN CHARLES BUCKNILL, F.R.S., & DR. DANIEL H. TUKE.
A MANUAL OF PSYCHOLOGICAL MEDICINE: containing the History, Nosology, Description, Statistics, Diagnosis, Pathology, and Treatment of Insanity. Second Edition. 8vo. cloth, 15s.

### DR. BUDD, F.R.S.
#### I.
ON DISEASES OF THE LIVER. Illustrated with Coloured Plates and Engravings on Wood. Third Edition. 8vo. cloth, 16s.
#### II.
ON THE ORGANIC DISEASES AND FUNCTIONAL DISORDERS OF THE STOMACH. 8vo. cloth, 9s.

### MR. CALLENDER, F.R.C.S.

FEMORAL RUPTURE : Anatomy of the Parts concerned. With Plates.
8vo. cloth, 4s.

### DR. JOHN M. CAMPLIN, F.L.S.

ON DIABETES, AND ITS SUCCESSFUL TREATMENT.
Third Edition, by Dr. Glover. Fcap. 8vo. cloth, 3s. 6d.

### MR. ROBERT B. CARTER, M.R.C.S.

**I.**

ON THE INFLUENCE OF EDUCATION AND TRAINING
IN PREVENTING DISEASES OF THE NERVOUS SYSTEM. Fcap. 8vo., 6s.

**II.**

THE PATHOLOGY AND TREATMENT OF HYSTERIA. Post
8vo. cloth, 4s. 6d.

### DR. CARPENTER, F.R.S.

**I.**

PRINCIPLES OF HUMAN PHYSIOLOGY. With numerous Illus-
trations on Steel and Wood. Sixth Edition. Edited by Mr. HENRY POWER. 8vo.
cloth, 26s.

**II.**

A MANUAL OF PHYSIOLOGY. With 252 Illustrations on Steel
and Wood.. Fourth Edition. Fcap. 8vo. cloth, 12s. 6d.

**III.**

THE MICROSCOPE AND ITS REVELATIONS. With more
than 400 Engravings on Steel and Wood. Fourth Edition. Fcap. 8vo. cloth, 12s. 6d.

### MR. JOSEPH PEEL CATLOW, M.R.C.S.

ON THE PRINCIPLES OF ÆSTHETIC MEDICINE ; or the
Natural Use of Sensation and Desire in the Maintenance of Health and the Treatment
of Disease. 8vo. cloth, 9s.

### DR. CHAMBERS.

**I.**

LECTURES, CHIEFLY CLINICAL. Fourth Edition. 8vo. cloth, 14s.

**II.**

THE INDIGESTIONS OR DISEASES OF THE DIGESTIVE
ORGANS FUNCTIONALLY TREATED. Second Edition. 8vo. cloth, 10s. 6d.

**III.**

SOME OF THE EFFECTS OF THE CLIMATE OF ITALY.
Crown 8vo. cloth, 4s. 6d.

### DR. CHANCE, M.B.

VIRCHOW'S CELLULAR PATHOLOGY, AS BASED UPON
PHYSIOLOGICAL AND PATHOLOGICAL HISTOLOGY. With 144 Engravings on Wood. 8vo. cloth, 16s.

### MR. H. T. CHAPMAN, F.R.C.S.

**I.**

THE TREATMENT OF OBSTINATE ULCERS AND CUTA-
NEOUS ERUPTIONS OF THE LEG WITHOUT CONFINEMENT. Third
Edition. Post 8vo. cloth, 3s. 6d.

**II.**

VARICOSE VEINS : their Nature, Consequences, and Treatment, Pallia-
tive and Curative. Second Edition. Post 8vo. cloth, 3s. 6d.

### MR. PYE HENRY CHAVASSE, F.R.C.S.

I.

## ADVICE TO A MOTHER ON THE MANAGEMENT OF
HER CHILDREN. Ninth Edition. Foolscap 8vo., 2s. 6d.

II.

## ADVICE TO A WIFE ON THE MANAGEMENT OF HER
OWN HEALTH. With an Introductory Chapter, especially addressed to a Young
Wife. Eighth Edition. Fcap. 8vo., 2s. 6d.

---

### MR. LE GROS CLARK, F.R.C.S.

## OUTLINES OF SURGERY ; being an Epitome of the Lectures on the
Principles and the Practice of Surgery delivered at St. Thomas's Hospital. Fcap. 8vo.
cloth, 5s.

---

### MR. JOHN CLAY, M.R.C.S.

## KIWISCH ON DISEASES OF THE OVARIES: Translated, by
permission, from the last German Edition of his Clinical Lectures on the Special Pathology and Treatment of the Diseases of Women. With Notes, and an Appendix on the
Operation of Ovariotomy. Royal 12mo. cloth, 16s.

---

### DR. COCKLE, M.D.

## ON INTRA-THORACIC CANCER. 8vo. 6s. 6d.

---

### MR. COLLIS, M.B.DUB., F.R.C.S.I.

## THE DIAGNOSIS AND TREATMENT OF CANCER AND
THE TUMOURS ANALOGOUS TO IT. With coloured Plates. 8vo. cloth, 14s.

---

### DR. CONOLLY.

## THE CONSTRUCTION AND GOVERNMENT OF LUNATIC
ASYLUMS AND HOSPITALS FOR THE INSANE. With Plans. Post 8vo.
cloth, 6s.

---

### MR. COOLEY.

COMPREHENSIVE SUPPLEMENT TO THE PHARMACOPŒIAS.

## THE CYCLOPÆDIA OF PRACTICAL RECEIPTS, PRO-
CESSES, AND COLLATERAL INFORMATION IN THE ARTS, MANU-
FACTURES, PROFESSIONS, AND TRADES, INCLUDING MEDICINE,
PHARMACY, AND DOMESTIC ECONOMY; designed as a General Book of
Reference for the Manufacturer, Tradesman, Amateur, and Heads of Families. Fourth
and greatly enlarged Edition, 8vo. cloth, 28s.

---

### MR. W. WHITE COOPER.

I.

## ON WOUNDS AND INJURIES OF THE EYE. Illustrated by
17 Coloured Figures and 41 Woodcuts. 8vo. cloth, 12s.

II.

## ON NEAR SIGHT, AGED SIGHT, IMPAIRED VISION,
AND THE MEANS OF ASSISTING SIGHT. With 31 Illustrations on Wood.
Second Edition. Fcap. 8vo. cloth, 7s. 6d.

### SIR ASTLEY COOPER, BART., F.R.S.
# ON THE STRUCTURE AND DISEASES OF THE TESTIS.
With 24 Plates. Second Edition. Royal 4to., 20s.

### MR. COOPER.
# A DICTIONARY OF PRACTICAL SURGERY AND ENCYCLO-
PÆDIA OF SURGICAL SCIENCE. New Edition, brought down to the present time. By SAMUEL A. LANE, F.R.C.S., assisted by various eminent Surgeons. Vol. I., 8vo. cloth, £1. 5s.

### MR. HOLMES COOTE, F.R.C.S.
# A REPORT ON SOME IMPORTANT POINTS IN THE
TREATMENT OF SYPHILIS. 8vo. cloth, 5s.

### DR. COTTON.
# PHTHISIS AND THE STETHOSCOPE; OR, THE PHYSICAL
SIGNS OF CONSUMPTION. Third Edition. Foolscap 8vo. cloth, 3s.

### MR. COULSON.
# ON DISEASES OF THE BLADDER AND PROSTATE GLAND.
New Edition, revised. *In Preparation.*

### MR. WALTER COULSON, F.R.C.S.
# STONE IN THE BLADDER: With Special Reference to its Prevention,
Early Symptoms, and Treatment by Lithotrity. 8vo. cloth, 6s.

### MR. WILLIAM CRAIG, L.F.P.S., GLASGOW.
# ON THE INFLUENCE OF VARIATIONS OF ELECTRIC
TENSION AS THE REMOTE CAUSE OF EPIDEMIC AND OTHER DISEASES. 8vo. cloth, 10s.

### MR. CURLING, F.R.S.
### I.
# OBSERVATIONS ON DISEASES OF THE RECTUM. Third
Edition. 8vo. cloth, 7s. 6d.
### II.
# A PRACTICAL TREATISE ON DISEASES OF THE TESTIS,
SPERMATIC CORD, AND SCROTUM. Third Edition, with Engravings. 8vo. cloth, 16s.

### DR. WILLIAM DALE, M.D.LOND.
# A COMPENDIUM OF PRACTICAL MEDICINE AND MORBID
ANATOMY. With Plates, 12mo. cloth, 7s.

### DR. DALRYMPLE, M.R.C.P., F.R.C.S.
# THE CLIMATE OF EGYPT: METEOROLOGICAL AND MEDI-
CAL OBSERVATIONS, with Practical Hints for Invalid Travellers. Post 8vo. cloth, 4s.

### MR. JOHN DALRYMPLE, F.R.S., F.R.C.S.
# PATHOLOGY OF THE HUMAN EYE. Complete in Nine Fasciculi:
imperial 4to., 20s. each; half-bound morocco, gilt tops, 9l. 15s.

### DR. HERBERT DAVIES.
# ON THE PHYSICAL DIAGNOSIS OF DISEASES OF THE
LUNGS AND HEART. Second Edition. Post 8vo. cloth, 8s.

### DR. DAVEY.

I.

THE GANGLIONIC NERVOUS SYSTEM: its Structure, Functions, and Diseases. 8vo. cloth, 9s.

II.

ON THE NATURE AND PROXIMATE CAUSE OF IN-SANITY. Post 8vo. cloth, 3s.

### DR. HENRY DAY, M.D., M.R.C.P.

CLINICAL HISTORIES; with Comments. 8vo. cloth, 7s. 6d.

### MR. DIXON.

A GUIDE TO THE PRACTICAL STUDY OF DISEASES OF THE EYE. Third Edition. Post 8vo. cloth, 9s.

### DR. DOBELL.

I.

DEMONSTRATIONS OF DISEASES IN THE CHEST, AND THEIR PHYSICAL DIAGNOSIS. With Coloured Plates. 8vo. cloth, 12s. 6d.

II.

LECTURES ON THE GERMS AND VESTIGES OF DISEASE, and on the Prevention of the Invasion and Fatality of Disease by Periodical Examinations. 8vo. cloth, 6s. 6d.

III.

ON TUBERCULOSIS: ITS NATURE, CAUSE, AND TREAT-MENT; with Notes on Pancreatic Juice. Second Edition. Crown 8vo. cloth, 3s. 6d.

IV.

LECTURES ON WINTER COUGH (CATARRH, BRONCHITIS, EMPHYSEMA, ASTHMA); with an Appendix on some Principles of Diet in Disease. Post 8vo. cloth, 5s. 6d.

V.

LECTURES ON THE TRUE FIRST STAGE OF CONSUMP-TION. Crown 8vo. cloth, 3s. 6d.

### DR. TOOGOOD DOWNING.

NEURALGIA: its various Forms, Pathology, and Treatment. THE JACKSONIAN PRIZE ESSAY FOR 1850. 8vo. cloth, 10s. 6d.

### DR. DRUITT, F.R.C.S.

THE SURGEON'S VADE-MECUM; with numerous Engravings on Wood. Ninth Edition. Foolscap 8vo. cloth, 12s. 6d.

### MR. DUNN, F.R.C.S.

PSYCHOLOGY—PHYSIOLOGICAL, 4s.; MEDICAL, 3s.

### SIR JAMES EYRE, M.D.

I.

THE STOMACH AND ITS DIFFICULTIES. Fifth Edition. Fcap. 8vo. cloth, 2s. 6d.

II.

PRACTICAL REMARKS ON SOME EXHAUSTING DIS-EASES. Second Edition. Post 8vo. cloth, 4s. 6d.

### DR. FAYRER, M.D., F.R.C.S.

CLINICAL SURGERY IN INDIA. With Engravings. 8vo. cloth, 16s.

### DR. FENWICK.

**I.**

## THE MORBID STATES OF THE STOMACH AND DUO-
DENUM, AND THEIR RELATIONS TO THE DISEASES OF OTHER ORGANS. With 10 Plates. 8vo. cloth, 12s.

**II.**

## ON SCROFULA AND CONSUMPTION. Clergyman's Sore Throat,
Catarrh, Croup, Bronchitis, Asthma. Fcap. 8vo., 2s. 6d.

### SIR WILLIAM FERGUSSON, BART., F.R.S.

**I.**

## A SYSTEM OF PRACTICAL SURGERY; with numerous Illus-
trations on Wood. Fourth Edition. Fcap. 8vo. cloth, 12s. 6d.

**II.**

## LECTURES ON THE PROGRESS OF ANATOMY AND
SURGERY DURING THE PRESENT CENTURY. With numerous Engravings. 8vo. cloth, 10s. 6d.

### SIR JOHN FIFE, F.R.C.S. AND MR. URQUHART.

## MANUAL OF THE TURKISH BATH. Heat a Mode of Cure and
a Source of Strength for Men and Animals. With Engravings. Post 8vo. cloth, 5s.

### MR. FLOWER, F.R.S., F.R.C.S.

## DIAGRAMS OF THE NERVES OF THE HUMAN BODY,
exhibiting their Origin, Divisions, and Connexions, with their Distribution to the various Regions of the Cutaneous Surface, and to all the Muscles. Folio, containing Six Plates, 14s.

### DR. BALTHAZAR FOSTER, M.D., M.R.C.P.

## THE USE OF THE SPHYGMOGRAPH IN THE INVESTI-
GATION OF DISEASE. With Engravings. 8vo. cloth, 2s. 6d.

### MR. FOWNES, PH.D., F.R.S.

**I.**

## A MANUAL OF CHEMISTRY; with 187 Illustrations on Wood.
Ninth Edition. Fcap. 8vo. cloth, 12s. 6d.
Edited by H. BENCE JONES, M.D., F.R.S., and A. W. HOFMANN, PH.D., F.R.S.

**II.**

## CHEMISTRY, AS EXEMPLIFYING THE WISDOM AND
BENEFICENCE OF GOD. Second Edition. Fcap. 8vo. cloth, 4s. 6d.

**III.**

## INTRODUCTION TO QUALITATIVE ANALYSIS. Post 8vo. cloth, 2s.

### DR. D. J. T. FRANCIS.

## CHANGE OF CLIMATE; considered as a Remedy in Dyspeptic, Pul-
monary, and other Chronic Affections; with an Account of the most Eligible Places of Residence for Invalids, at different Seasons of the Year. Post 8vo. cloth, 8s. 6d.

### DR. W. FRAZER.

## ELEMENTS OF MATERIA MEDICA; containing the Chemistry
and Natural History of Drugs—their Effects, Doses, and Adulterations. Second Edition. 8vo. cloth, 10s. 6d.

### PROFESSOR FRESENIUS.

## A SYSTEM OF INSTRUCTION IN CHEMICAL ANALYSIS,
Edited by LLOYD BULLOCK, F.C.S.
QUALITATIVE. Sixth Edition, with Coloured Plate illustrating Spectrum Analysis. 8vo. cloth, 10s. 6d.——QUANTITATIVE. Fourth Edition. 8vo. cloth, 18s.

### DR. FULLER.

#### I.
## ON DISEASES OF THE LUNGS AND AIR PASSAGES.
Second Edition. 8vo. cloth, 12s. 6d.

#### II.
## ON DISEASES OF THE HEART AND GREAT VESSELS.
8vo. cloth, 7s. 6d.

#### III.
## ON RHEUMATISM, RHEUMATIC GOUT, AND SCIATICA:
their Pathology, Symptoms, and Treatment.    Third Edition.    8vo. cloth, 12s. 6d.

### MR. GALLOWAY.

#### I.
## THE FIRST STEP IN CHEMISTRY. With numerous Engravings.
Fourth Edition. Fcap. 8vo. cloth, 6s. 6d.    II.

## A KEY TO THE EXERCISES CONTAINED IN ABOVE. Fcap.
8vo., 2s. 6d.    III.

## THE SECOND STEP IN CHEMISTRY; or, the Student's Guide to
the Higher Branches of the Science.    With Engravings.    8vo. cloth, 10s.

#### IV.
## A MANUAL OF QUALITATIVE ANALYSIS. Fourth Edition.
Post 8vo. cloth, 6s. 6d.    V.

## CHEMICAL TABLES. On Five Large Sheets, for School and Lecture
Rooms.    Second Edition.    4s. 6d.

### MR. J. SAMPSON GAMGEE.
## HISTORY OF A SUCCESSFUL CASE OF AMPUTATION AT
THE HIP-JOINT (the limb 48-in. in circumference, 99 pounds weight).    With 4
Photographs.    4to cloth, 10s. 6d.

### MR. F. J. GANT, F.R.C.S.

#### I.
## THE PRINCIPLES OF SURGERY: Clinical, Medical, and Opera-
tive.    With Engravings.    8vo. cloth, 18s.

#### II.
## THE IRRITABLE BLADDER: its Causes and Curative Treatment.
Second Edition, enlarged.    Crown 8vo. cloth, 5s.

### SIR DUNCAN GIBB, BART., M.D.

#### I.
## ON DISEASES OF THE THROAT AND WINDPIPE, as
reflected by the Laryngoscope.    Second Edition.    With 116 Engravings.    Post 8vo.
cloth, 10s. 6d.    II.

## THE LARYNGOSCOPE IN DISEASES OF THE THROAT,
with a Chapter on RHINOSCOPY.    Second Edition, enlarged, with Engravings.    Crown
8vo., cloth, 5s.

### MRS. GODFREY.
## ON THE NATURE, PREVENTION, TREATMENT, AND CURE
OF SPINAL CURVATURES and DEFORMITIES of the CHEST and LIMBS,
without ARTIFICIAL SUPPORTS or any MECHANICAL APPLIANCES.
Third Edition, Revised and Enlarged.    8vo. cloth 5s.

### DR. GORDON, M.D., C.B.

#### I.
## ARMY HYGIENE. 8vo. cloth, 20s.

#### II.
## CHINA, FROM A MEDICAL POINT OF VIEW; IN 1860
AND 1861; With a Chapter on Nagasaki as a Sanatarium.    8vo. cloth, 10s. 6d.

### DR. GAIRDNER.

ON GOUT; its History, its Causes, and its Cure. Fourth Edition. Post 8vo. cloth, 8s. 6d.

### DR. GRANVILLE, F.R.S.

I.

THE MINERAL SPRINGS OF VICHY: their Efficacy in the Treatment of Gout, Indigestion, Gravel, &c. 8vo. cloth, 3s.

II.

ON SUDDEN DEATH. Post 8vo., 2s. 6d.

### DR. GRAVES M.D., F.R.S.

STUDIES IN PHYSIOLOGY AND MEDICINE. Edited by Dr. Stokes. With Portrait and Memoir. 8vo. cloth, 14s.

### DR. S. C. GRIFFITH, M.D.

ON DERMATOLOGY AND THE TREATMENT OF SKIN DISEASES BY MEANS OF HERBS, IN PLACE OF ARSENIC AND MERCURY. Fcap. 8vo. cloth, 3s.

### MR. GRIFFITHS.

CHEMISTRY OF THE FOUR SEASONS—Spring, Summer, Autumn, Winter. Illustrated with Engravings on Wood. Second Edition. Foolscap 8vo. cloth, 7s. 6d.

### DR. GULLY.

THE SIMPLE TREATMENT OF DISEASE; deduced from the Methods of Expectancy and Revulsion. 18mo. cloth, 4s.

### DR. GUY AND DR. JOHN HARLEY.

HOOPER'S PHYSICIAN'S VADE-MECUM; OR, MANUAL OF THE PRINCIPLES AND PRACTICE OF PHYSIC. Seventh Edition, considerably enlarged, and rewritten. Foolscap 8vo. cloth, 12s. 6d.

GUY'S HOSPITAL REPORTS. Third Series. Vol. XIII., 8vo. 7s. 6d.

### DR. HABERSHON, F.R.C.P.

I.

ON DISEASES OF THE ABDOMEN, comprising those of the Stomach and other Parts of the Alimentary Canal, Œsophagus, Stomach, Cæcum, Intestines, and Peritoneum. Second Edition, with Plates. 8vo. cloth, 14s.

II.

ON THE INJURIOUS EFFECTS OF MERCURY IN THE TREATMENT OF DISEASE. Post 8vo. cloth, 3s. 6d.

### DR. C. RADCLYFFE HALL.

TORQUAY IN ITS MEDICAL ASPECT AS A RESORT FOR PULMONARY INVALIDS. Post 8vo. cloth, 5s.

### DR. MARSHALL HALL, F.R.S.

I.

PRONE AND POSTURAL RESPIRATION IN DROWNING AND OTHER FORMS OF APNŒA OR SUSPENDED RESPIRATION. Post 8vo. cloth, 5s.

II.

PRACTICAL OBSERVATIONS AND SUGGESTIONS IN MEDICINE. Second Series. Post 8vo. cloth, 8s. 6d.

### MR. HARDWICH.

## A MANUAL OF PHOTOGRAPHIC CHEMISTRY. With
Engravings. Seventh Edition. Foolscap 8vo. cloth, 7s. 6d.

### DR. J. BOWER HARRISON, M.D., M.R.C.P.

#### I.

## LETTERS TO A YOUNG PRACTITIONER ON THE DIS-
EASES OF CHILDREN. Foolscap 8vo. cloth, 3s.

#### II.

## ON THE CONTAMINATION OF WATER BY THE POISON
OF LEAD, and its Effects on the Human Body. Foolscap 8vo. cloth, 3s. 6d.

### DR. HARTWIG.

#### I.

## ON SEA BATHING AND SEA AIR. Second Edition. Fcap.
8vo., 2s. 6d.

#### II.

## ON THE PHYSICAL EDUCATION OF CHILDREN. Fcap.
8vo., 2s. 6d.

### DR. A. H. HASSALL.

#### I.

## THE URINE, IN HEALTH AND DISEASE; being an Ex-
planation of the Composition of the Urine, and of the Pathology and Treatment of
Urinary and Renal Disorders. Second Edition. With 79 Engravings (23 Coloured).
Post 8vo. cloth, 12s. 6d.

#### II.

## THE MICROSCOPIC ANATOMY OF THE HUMAN BODY,
IN HEALTH AND DISEASE. Illustrated with Several Hundred Drawings in
Colour. Two vols. 8vo. cloth, £1. 10s.

### MR. ALFRED HAVILAND, M.R.C.S.

## CLIMATE, WEATHER, AND DISEASE; being a Sketch of the
Opinions of the most celebrated Ancient and Modern Writers with regard to the Influence
of Climate and Weather in producing Disease. With Four coloured Engravings. 8vo.
cloth, 7s.

### DR. HEADLAND, M.D., F.R.C.P.

#### I.

## ON THE ACTION OF MEDICINES IN THE SYSTEM.
Fourth Edition. 8vo. cloth, 14s.

#### II.

## A MEDICAL HANDBOOK; comprehending such Information on Medical
and Sanitary Subjects as is desirable in Educated Persons. Second Thousand. Foolscap
8vo. cloth, 5s.

### DR. HEALE.

#### I.

## A TREATISE ON THE PHYSIOLOGICAL ANATOMY OF
THE LUNGS. With Engravings. 8vo. cloth, 8s.

#### II.

## A TREATISE ON VITAL CAUSES. 8vo. cloth, 9s.

### MR. CHRISTOPHER HEATH, F.R.C.S.

I.

PRACTICAL ANATOMY: a Manual of Dissections. With numerous Engravings. Fcap. 8vo. cloth, 10s. 6d.

II.

A MANUAL OF MINOR SURGERY AND BANDAGING, FOR THE USE OF HOUSE-SURGEONS, DRESSERS, AND JUNIOR PRACTITIONERS. With Illustrations. Third Edition. Fcap. 8vo. cloth, 5s.

### MR. HIGGINBOTTOM, F.R.S., F.R.C.S.E.

A PRACTICAL ESSAY ON THE USE OF THE NITRATE OF SILVER IN THE TREATMENT OF INFLAMMATION, WOUNDS, AND ULCERS. Third Edition, 8vo. cloth, 6s.

### DR. HINDS.

THE HARMONIES OF PHYSICAL SCIENCE IN RELATION TO THE HIGHER SENTIMENTS; with Observations on Medical Studies, and on the Moral and Scientific Relations of Medical Life. Post 8vo. cloth, 4s.

### MR. J. A. HINGESTON, M.R.C.S.

TOPICS OF THE DAY, MEDICAL, SOCIAL, AND SCIENTIFIC. Crown 8vo. cloth, 7s. 6d.

### DR. HODGES.

THE NATURE, PATHOLOGY, AND TREATMENT OF PUERPERAL CONVULSIONS. Crown 8vo. cloth, 3s.

### DR. DECIMUS HODGSON.

THE PROSTATE GLAND, AND ITS ENLARGEMENT IN OLD AGE. With 12 Plates. Royal 8vo. cloth, 6s.

### MR. JABEZ HOGG.

A MANUAL OF OPHTHALMOSCOPIC SURGERY; being a Practical Treatise on the Use of the Ophthalmoscope in Diseases of the Eye. Third Edition. With Coloured Plates. 8vo. cloth, 10s. 6d.

### MR. LUTHER HOLDEN, F.R.C.S.

I.

HUMAN OSTEOLOGY: with Plates, showing the Attachments of the Muscles. Third Edition. 8vo. cloth, 16s.

II.

A MANUAL OF THE DISSECTION OF THE HUMAN BODY. With Engravings on Wood. Second Edition. 8vo. cloth, 16s.

### MR. BARNARD HOLT, F.R.C.S.

ON THE IMMEDIATE TREATMENT OF STRICTURE OF THE URETHRA. Third Edition, Enlarged. 8vo. cloth, 6s.

**DR. W. CHARLES HOOD.**

SUGGESTIONS FOR THE FUTURE PROVISION OF CRIMI-
NAL LUNATICS. 8vo. cloth, 5s. 6d.

**DR. P. HOOD.**

THE SUCCESSFUL TREATMENT OF SCARLET FEVER;
also, OBSERVATIONS ON THE PATHOLOGY AND TREATMENT OF
CROWING INSPIRATIONS OF INFANTS. Post 8vo. cloth, 5s.

**MR. JOHN HORSLEY.**

A CATECHISM OF CHEMICAL PHILOSOPHY; being a Familiar
Exposition of the Principles of Chemistry and Physics. With Engravings on Wood.
Designed for the Use of Schools and Private Teachers. Post 8vo. cloth, 6s. 6d.

**DR. JAMES A. HORTON, M.D.**

PHYSICAL AND MEDICAL CLIMATE AND METEOROLOGY
OF THE WEST COAST OF AFRICA. 8vo. cloth, 10s.

**MR. LUKE HOWARD, F.R.S.**

ESSAY ON THE MODIFICATIONS OF CLOUDS. Third Edition,
by W. D. and E. HOWARD. With 6 Lithographic Plates, from Pictures by Kenyon.
4to. cloth, 10s. 6d.

**DR. HAMILTON HOWE, M.D.**

A THEORETICAL INQUIRY INTO THE PHYSICAL CAUSE
OF EPIDEMIC DISEASES. Accompanied with Tables. 8vo. cloth, 7s.

**DR. HUFELAND.**

THE ART OF PROLONGING LIFE. Second Edition. Edited
by ERASMUS WILSON, F.R.S. Foolscap 8vo., 2s. 6d.

**MR. W. CURTIS HUGMAN, F.R.C.S.**

ON HIP-JOINT DISEASE; with reference especially to Treatment
by Mechanical Means for the Relief of Contraction and Deformity of the Affected Limb.
With Plates. Re-issue, enlarged. 8vo. cloth, 3s. 6d.

**MR. HULKE, F.R.C.S.**

A PRACTICAL TREATISE ON THE USE OF THE
OPHTHALMOSCOPE. Being the Jacksonian Prize Essay for 1859. Royal 8vo.
cloth, 8s.

**DR. HENRY HUNT.**

ON HEARTBURN AND INDIGESTION. 8vo. cloth, 5s.

**MR. G. Y. HUNTER, M.R.C.S.**

BODY AND MIND: the Nervous System and its Derangements.
Fcap. 8vo. cloth, 3s. 6d.

**PROFESSOR HUXLEY, F.R.S.**

LECTURES ON THE ELEMENTS OF COMPARATIVE
ANATOMY.—ON CLASSIFICATON AND THE SKULL. With 111 Illustrations. 8vo. cloth, 10s. 6d.

### MR. JONATHAN HUTCHINSON, F.R.C.S.

## A CLINICAL MEMOIR ON CERTAIN DISEASES OF THE
EYE AND EAR, CONSEQUENT ON INHERITED SYPHILIS; with an appended Chapter of Commentaries on the Transmission of Syphilis from Parent to Offspring, and its more remote Consequences.  With Plates and Woodcuts, 8vo. cloth, 9s.

### DR. INMAN, M.R.C.P.
#### I.

## ON MYALGIA: ITS NATURE, CAUSES, AND TREATMENT;
being a Treatise on Painful and other Affections of the Muscular System.  Second Edition.  8vo. cloth, 9s.

#### II.

## FOUNDATION FOR A NEW THEORY AND PRACTICE
OF MEDICINE.  Second Edition.  Crown 8vo. cloth, 10s.

### DR. JAGO, M.D.OXON., A.B.CANTAB.

## ENTOPTICS, WITH ITS USES IN PHYSIOLOGY AND
MEDICINE.  With 54 Engravings.  Crown 8vo. cloth, 5s.

### MR. J. H. JAMES, F.R.C.S.
#### I.

## PRACTICAL OBSERVATIONS ON THE OPERATIONS FOR
STRANGULATED HERNIA.  8vo. cloth, 5s.

#### II.

## ON THE DISTINCTIVE CHARACTERS OF EXTERNAL
INFLAMMATIONS, AND ON INFLAMMATORY OR SYMPATHETIC FEVER.  8vo. cloth, 5s.

### DR. PROSSER JAMES, M.D.

## SORE-THROAT: ITS NATURE, VARIETIES, AND TREAT-
MENT; including the Use of the LARYNGOSCOPE as an Aid to Diagnosis.  Second Edition, with numerous Engravings.  Post 8vo. cloth, 5s.

### DR. JENCKEN, M.D., M.R.C.P.

## THE CHOLERA: ITS ORIGIN, IDIOSYNCRACY, AND
TREATMENT.  Fcap. 8vo. cloth, 2s. 6d.

### DR. HANDFIELD JONES, M.B., F.R.C.P.

## CLINICAL OBSERVATIONS ON FUNCTIONAL NERVOUS
DISORDERS.  Post 8vo. cloth, 10s. 6d.

### DR. H. BENCE JONES, M.D., F.R.S.

## LECTURES ON SOME OF THE APPLICATIONS OF
CHEMISTRY AND MECHANICS TO PATHOLOGY AND THERA-PEUTICS.  8vo. cloth, 12s.

### DR. HANDFIELD JONES, F.R.S., & DR. EDWARD H. SIEVEKING.

## A MANUAL OF PATHOLOGICAL ANATOMY.  Illustrated with
numerous Engravings on Wood.  Foolscap 8vo. cloth, 12s. 6d.

### DR. JAMES JONES, M.D., M.R.C.P.

## ON THE USE OF PERCHLORIDE OF IRON AND OTHER
CHALYBEATE SALTS IN THE TREATMENT OF CONSUMPTION.  Crown 8vo. cloth, 3s. 6d.

**MR. WHARTON JONES, F.R.S.**

I.

## A MANUAL OF THE PRINCIPLES AND PRACTICE OF

OPHTHALMIC MEDICINE AND SURGERY; with Nine Coloured Plates and 173 Wood Engravings. Third Edition, thoroughly revised. Foolscap 8vo. cloth, 12s. 6d.

II.

## THE WISDOM AND BENEFICENCE OF THE ALMIGHTY,

AS DISPLAYED IN THE SENSE OF VISION. Actonian Prize Essay. With Illustrations on Steel and Wood. Foolscap 8vo. cloth, 4s. 6d.

III.

## DEFECTS OF SIGHT AND HEARING: their Nature, Causes, Pre-

vention, and General Management. Second Edition, with Engravings. Fcap. 8vo. 2s. 6d.

IV.

## A CATECHISM OF THE MEDICINE AND SURGERY OF

THE EYE AND EAR. For the Clinical Use of Hospital Students. Fcap. 8vo. 2s. 6d.

V.

## A CATECHISM OF THE PHYSIOLOGY AND PHILOSOPHY

OF BODY, SENSE, AND MIND. For Use in Schools and Colleges. Fcap. 8vo., 2s. 6d.

---

**MR. FURNEAUX JORDAN, M.R.C.S.**

## AN INTRODUCTION TO CLINICAL SURGERY; WITH A

Method of Investigating and Reporting Surgical Cases. Fcap. 8vo. cloth, 5s.

---

**MR. JUDD.**

## A PRACTICAL TREATISE ON URETHRITIS AND SYPHI-

LIS: including Observations on the Power of the Menstruous Fluid, and of the Discharge from Leucorrhœa and Sores to produce Urethritis: with a variety of Examples, Experiments, Remedies, and Cures. 8vo. cloth, £1. 5s.

---

**DR. LAENNEC.**

## A MANUAL OF AUSCULTATION AND PERCUSSION. Trans-

lated and Edited by J. B. SHARPE, M.R.C.S. 3s.

---

**DR. LANE, M.A.**

## HYDROPATHY; OR, HYGIENIC MEDICINE. An Explanatory

Essay. Second Edition. Post 8vo. cloth, 5s.

---

**SIR WM. LAWRENCE, BART., F.R.S.**

I.

## LECTURES ON SURGERY. 8vo. cloth, 16s.

II.

## A TREATISE ON RUPTURES. The Fifth Edition, considerably

enlarged. 8vo. cloth, 16s.

---

**DR. LEARED, M.R.C.P.**

## IMPERFECT DIGESTION: ITS CAUSES AND TREATMENT.

Fourth Edition. Foolscap 8vo. cloth, 4s.

### DR. EDWIN LEE.

**I.**

## THE EFFECT OF CLIMATE ON TUBERCULOUS DISEASE,
with Notices of the chief Foreign Places of Winter Resort. Small 8vo. cloth, 4s. 6d.

**II.**

## THE WATERING PLACES OF ENGLAND, CONSIDERED
with Reference to their Medical Topography. Fourth Edition. Fcap. 8vo. cloth, 7s. 6d.

**III.**

## THE PRINCIPAL BATHS OF FRANCE. Fourth Edition.
Fcap. 8vo. cloth, 3s. 6d.

**IV.**

## THE BATHS OF GERMANY. Fourth Edition. Post 8vo. cloth, 7s.

**V.**

## THE BATHS OF SWITZERLAND. 12mo. cloth, 3s. 6d.

**VI.**

## HOMŒOPATHY AND HYDROPATHY IMPARTIALLY AP-
PRECIATED. Fourth Edition. Post 8vo. cloth, 3s.

---

### MR. HENRY LEE, F.R.C.S.

**I.**

## ON SYPHILIS. Second Edition. With Coloured Plates. 8vo. cloth, 10s.

**II.**

## ON DISEASES OF THE VEINS, HÆMORRHOIDAL TUMOURS,
AND OTHER AFFECTIONS OF THE RECTUM. Second Edition. 8vo. cloth, 8s.

---

### DR. ROBERT LEE, F.R.S.

**I.**

## CONSULTATIONS IN MIDWIFERY. Foolscap 8vo. cloth, 4s. 6d.

**II.**

## A TREATISE ON THE SPECULUM; with Three Hundred Cases.
8vo. cloth, 4s. 6d.

**III.**

## CLINICAL REPORTS OF OVARIAN AND UTERINE DIS-
EASES, with Commentaries. Foolscap 8vo. cloth, 6s. 6d.

**IV.**

## CLINICAL MIDWIFERY: comprising the Histories of 545 Cases of
Difficult, Preternatural, and Complicated Labour, with Commentaries. Second Edition.
Foolscap 8vo. cloth, 5s.

---

### DR. LEISHMAN, M.D., F.F.P.S.

## THE MECHANISM OF PARTURITION: An Essay, Historical and
Critical. With Engravings. 8vo. cloth, 5s.

---

### MR. LISTON, F.R.S.

## PRACTICAL SURGERY. Fourth Edition. 8vo. cloth, 22s.

---

### MR. H. W. LOBB, L.S.A., M.R.C.S.E.

## ON SOME OF THE MORE OBSCURE FORMS OF NERVOUS
AFFECTIONS, THEIR PATHOLOGY AND TREATMENT. Re-issue,
with the Chapter on Galvanism entirely Re-written. With Engravings. 8vo. cloth, 8s.

---

### DR. LOGAN, M.D., M.R.C.P.LOND.

## ON OBSTINATE DISEASES OF THE SKIN. Fcap. 8vo. cloth, 2s. 6d.

**LONDON HOSPITAL.**

## CLINICAL LECTURES AND REPORTS BY THE MEDICAL
AND SURGICAL STAFF. With Illustrations. Vols. I. to IV. 8vo. cloth, 7s. 6d.

**LONDON MEDICAL SOCIETY OF OBSERVATION.**

## WHAT TO OBSERVE AT THE BED-SIDE, AND AFTER
DEATH. Published by Authority. Second Edition. Foolscap 8vo. cloth, 4s. 6d.

**MR. HENRY LOWNDES, M.R.C.S.**

## AN ESSAY ON THE MAINTENANCE OF HEALTH. Fcap.
8vo. cloth, 2s. 6d.

**MR. M'CLELLAND, F.L.S., F.G.S.**

## THE MEDICAL TOPOGRAPHY, OR CLIMATE AND SOILS,
OF BENGAL AND THE N. W. PROVINCES. Post 8vo. cloth, 4s. 6d.

**DR. MACLACHLAN, M.D., F.R.C.P.L.**

## THE DISEASES AND INFIRMITIES OF ADVANCED LIFE.
8vo. cloth, 16s.

**DR. A. C. MACLEOD, M.R.C.P.LOND.**

## ACHOLIC DISEASES ; comprising Jaundice, Diarrhœa, Dysentery,
and Cholera. Post 8vo. cloth, 5s. 6d.

**DR. GEORGE H. B. MACLEOD, F.R.C.S.E.**

I.

## OUTLINES OF SURGICAL DIAGNOSIS. 8vo. cloth, 12s. 6d.

II.

## NOTES ON THE SURGERY OF THE CRIMEAN WAR; with
REMARKS on GUN-SHOT WOUNDS. 8vo. cloth, 10s. 6d.

**DR. WM. MACLEOD, F.R.C.P.EDIN.**

## THE THEORY OF THE TREATMENT OF DISEASE ADOPTED
AT BEN RHYDDING. Fcap. 8vo. cloth, 2s. 6d.

**MR. JOSEPH MACLISE, F.R.C.S.**

I.

## SURGICAL ANATOMY. A Series of Dissections, illustrating the Prin-
cipal Regions of the Human Body. Second Edition, folio, cloth, £3. 12s.; half-morocco,
£4. 4s.

II.

## ON DISLOCATIONS AND FRACTURES. This Work is Uniform
with "Surgical Anatomy;" folio, cloth, £2. 10s.; half-morocco, £2. 17s.

**MR. MACNAMARA.**

## ON DISEASES OF THE EYE; referring principally to those Affections
requiring the aid of the Ophthalmoscope for their Diagnosis. With coloured plates.
8vo. cloth, 10s. 6d.

**DR. MCNICOLL, M.R.C.P.**

## A HAND-BOOK FOR SOUTHPORT, MEDICAL & GENERAL;
with Copious Notices of the Natural History of the District. Second Edition. Post 8vo.
cloth, 3s. 6d.

**DR. MARCET, F.R.S.**

## ON CHRONIC ALCOHOLIC INTOXICATION; with an INQUIRY
INTO THE INFLUENCE OF THE ABUSE OF ALCOHOL AS A PRE-
DISPOSING CAUSE OF DISEASE. Second Edition, much enlarged. Foolscap
8vo. cloth, 4s. 6d.

### DR. J. MACPHERSON, M.D.

CHOLERA IN ITS HOME; with a Sketch of the Pathology and Treatment of the Disease. Crown 8vo. cloth, 5s.

### DR. MARKHAM.

**I.**

DISEASES OF THE HEART: THEIR PATHOLOGY, DIAGNOSIS, AND TREATMENT. Second Edition. Post 8vo. cloth, 6s.

**II.**

SKODA ON AUSCULTATION AND PERCUSSION. Post 8vo. cloth, 6s.

**III.**

BLEEDING AND CHANGE IN TYPE OF DISEASES. Gulstonian Lectures for 1864. Crown 8vo. 2s. 6d.

### SIR RANALD MARTIN, K.C.B., F.R.S.

INFLUENCE OF TROPICAL CLIMATES IN PRODUCING THE ACUTE ENDEMIC DISEASES OF EUROPEANS; including Practical Observations on their Chronic Sequelæ under the Influences of the Climate of Europe. Second Edition, much enlarged. 8vo. cloth, 20s.

### MR. C. F. MAUNDER, F.R.C.S.

OPERATIVE SURGERY. With 158 Engravings. Post 8vo. 6s.

### DR. MAYNE, M.D., LL.D.

**I.**

AN EXPOSITORY LEXICON OF THE TERMS, ANCIENT AND MODERN, IN MEDICAL AND GENERAL SCIENCE. 8vo. cloth, £2. 10s.

**II.**

A MEDICAL VOCABULARY; or, an Explanation of all Names, Synonymes, Terms, and Phrases used in Medicine and the relative branches of Medical Science. Second Edition. Fcap. 8vo. cloth, 8s. 6d.

### DR. MERYON, M.D., F.R.C.P.

PATHOLOGICAL AND PRACTICAL RESEARCHES ON THE VARIOUS FORMS OF PARALYSIS. 8vo. cloth, 6s.

### DR. MILLINGEN.

ON THE TREATMENT AND MANAGEMENT OF THE INSANE; with Considerations on Public and Private Lunatic Asylums. 18mo. cloth, 4s. 6d.

### DR. W. J. MOORE, M.D.

**I.**

HEALTH IN THE TROPICS; or, Sanitary Art applied to Europeans in India. 8vo. cloth, 9s.

**II.**

A MANUAL OF THE DISEASES OF INDIA. Fcap. 8vo. cloth, 5s.

### DR. JAMES MORRIS, M.D.LOND.

**I.**

GERMINAL MATTER AND THE CONTACT THEORY: An Essay on the Morbid Poisons. Second Edition. Crown 8vo. cloth, 4s. 6d.

**II.**

IRRITABILITY: Popular and Practical Sketches of Common Morbid States and Conditions bordering on Disease; with Hints for Management, Alleviation, and Cure. Crown 8vo. cloth, 4s. 6d.

**PROFESSOR MULDER, UTRECHT.**

THE CHEMISTRY OF WINE. Edited by H. BENCE JONES, M.D., F.R.S. Fcap. 8vo. cloth, 6s.

**DR. W. MURRAY, M.D., M.R.C.P.**

EMOTIONAL DISORDERS OF THE SYMPATHETIC SYSTEM OF NERVES. Crown 8vo. cloth, 3s. 6d.

**DR. MUSHET, M.B., M.R.C.P.**

ON APOPLEXY, AND ALLIED AFFECTIONS OF THE BRAIN. 8vo. cloth, 7s.

**MR. NAYLER, F.R.C.S.**

ON THE DISEASES OF THE SKIN. With Plates. 8vo. cloth, 10s. 6d.

**DR. BIRKBECK NEVINS.**

THE PRESCRIBER'S ANALYSIS OF THE BRITISH PHARMACOPEIA of 1867. 32mo. cloth, 3s. 6d.

**DR. THOS. NICHOLSON, M.D.**

ON YELLOW FEVER; comprising the History of that Disease as it appeared in the Island of Antigua. Fcap. 8vo. cloth, 2s. 6d.

**DR. NOAD, PH.D., F.R.S.**

THE INDUCTION COIL, being a Popular Explanation of the Electrical Principles on which it is constructed. Third Edition. With Engravings. Fcap. 8vo. cloth, 3s.

**DR. NOBLE.**

THE HUMAN MIND IN ITS RELATIONS WITH THE BRAIN AND NERVOUS SYSTEM. Post 8vo. cloth, 4s. 6d.

**MR. NUNNELEY, F.R.C.S.E.**

I.

ON THE ORGANS OF VISION: THEIR ANATOMY AND PHYSIOLOGY. With Plates, 8vo. cloth, 15s.

II.

A TREATISE ON THE NATURE, CAUSES, AND TREATMENT OF ERYSIPELAS. 8vo. cloth, 10s. 6d.

**DR. OPPERT, M.D.**

I.

HOSPITALS, INFIRMARIES, AND DISPENSARIES; their Construction, Interior Arrangement, and Management, with Descriptions of existing Institutions. With 58 Engravings. Royal 8vo. cloth, 10s. 6d.

II.

VISCERAL AND HEREDITARY SYPHILIS. 8vo. cloth, 5s.

**MR. LANGSTON PARKER.**

THE MODERN TREATMENT OF SYPHILITIC DISEASES, both Primary and Secondary; comprising the Treatment of Constitutional and Confirmed Syphilis, by a safe and successful Method. Fourth Edition, 8vo. cloth, 10s.

**DR. PARKES, F.R.S., F.R.C.P.**

I.

A MANUAL OF PRACTICAL HYGIENE; intended especially for the Medical Officers of the Army. With Plates and Woodcuts. 2nd Edition, 8vo. cloth, 16s.

II.

THE URINE: ITS COMPOSITION IN HEALTH AND DISEASE, AND UNDER THE ACTION OF REMEDIES. 8vo. cloth, 12s.

### DR. PARKIN, M.D., F.R.C.S.

#### I.

## THE ANTIDOTAL TREATMENT AND PREVENTION OF

THE EPIDEMIC CHOLERA. Third Edition. 8vo. cloth, 7s. 6d.

#### II.

## THE CAUSATION AND PREVENTION OF DISEASE; with

the Laws regulating the Extrication of Malaria from the Surface, and its Diffusion in the surrounding Air. 8vo. cloth, 5s.

### MR. JAMES PART, F.R.C.S.

## THE MEDICAL AND SURGICAL POCKET CASE BOOK,

for the Registration of important Cases in Private Practice, and to assist the Student of Hospital Practice. Second Edition. 2s. 6d.

### DR. PATTERSON, M.D.

## EGYPT AND THE NILE AS A WINTER RESORT FOR

PULMONARY AND OTHER INVALIDS. Fcap. 8vo. cloth, 3s.

### DR. PAVY, M.D., F.R.S., F.R.C.P.

#### I.

## DIABETES : RESEARCHES ON ITS NATURE AND TREAT-

MENT. 8vo. cloth, 8s. 6d.

#### II.

## DIGESTION : ITS DISORDERS AND THEIR TREATMENT.

8vo. cloth, 8s. 6d.

### DR. PEACOCK, M.D., F.R.C.P.

#### I.

## ON MALFORMATIONS OF THE HUMAN HEART. With

Original Cases and Illustrations. Second Edition. With 8 Plates. 8vo. cloth, 10s.

#### II.

## ON SOME OF THE CAUSES AND EFFECTS OF VALVULAR

DISEASE OF THE HEART. With Engravings. 8vo. cloth, 5s.

### DR. W. H. PEARSE, M.D.EDIN.

## NOTES ON HEALTH IN CALCUTTA AND BRITISH

EMIGRANT SHIPS, including Ventilation, Diet, and Disease. Fcap. 8vo. 2s.

### DR. PEET, M.D., F.R.C.P.

## THE PRINCIPLES AND PRACTICE OF MEDICINE

Designed chiefly for Students of Indian Medical Colleges. 8vo. cloth, 16s.

### DR. PEREIRA, F.R.S.

## SELECTA E PRÆSCRIPTIS. Fourteenth Edition. 24mo. cloth, 5s.

### DR. PICKFORD.

## HYGIENE; or, Health as Depending upon the Conditions of the Atmo-

sphere, Food and Drinks, Motion and Rest, Sleep and Wakefulness, Secretions, Excretions, and Retentions, Mental Emotions, Clothing, Bathing, &c. Vol. I. 8vo. cloth, 9s.

'PROFESSOR PIRRIE, F.R.S.E.

## THE PRINCIPLES AND PRACTICE OF SURGERY. With
numerous Engravings on Wood. Second Edition. 8vo. cloth, 24s.

PROFESSOR PIRRIE & DR. KEITH.

ACUPRESSURE: an excellent Method of arresting Surgical Hæmorrhage
and of accelerating the healing of Wounds. With Engravings. 8vo. cloth, 5s.

DR. PIRRIE, M.D.

## ON HAY ASTHMA, AND THE AFFECTION TERMED
HAY FEVER. Fcap. 8vo. cloth, 2s. 6d.

PROFESSORS PLATTNER & MUSPRATT.

## THE USE OF THE BLOWPIPE IN THE EXAMINATION OF
MINERALS, ORES, AND OTHER METALLIC COMBINATIONS. Illustrated
by numerous Engravings on Wood. Third Edition. 8vo. cloth, 10s. 6d.

MR. HENRY POWER, F.R.C.S., M.B.LOND.

## ILLUSTRATIONS OF SOME OF THE PRINCIPAL DISEASES
OF THE EYE: With an Account of their Symptoms, Pathology and Treatment.
Twelve Coloured Plates. 8vo. cloth, 20s.

DR. HENRY F. A. PRATT, M.D., M.R.C.P.

I.

## THE GENEALOGY OF CREATION, newly Translated from the
Unpointed Hebrew Text of the Book of Genesis, showing the General Scientific Accuracy
of the Cosmogony of Moses and the Philosophy of Creation. 8vo. cloth, 14s.

II.

## ON ECCENTRIC AND CENTRIC FORCE: A New Theory of
Projection. With Engravings. 8vo. cloth, 10s.

III.

## ON ORBITAL MOTION: The Outlines of a System of Physical
Astronomy. With Diagrams. 8vo. cloth, 7s. 6d.

IV.

## ASTRONOMICAL INVESTIGATIONS. The Cosmical Relations of
the Revolution of the Lunar Apsides. Oceanic Tides. With Engravings. 8vo. cloth, 5s.

V.

## THE ORACLES OF GOD: An Attempt at a Re-interpretation. Part I.
The Revealed Cosmos. 8vo. cloth, 10s.

## THE PRESCRIBER'S PHARMACOPŒIA; containing all the Medi-
cines in the British Pharmacopœia, arranged in Classes according to their Action, with
their Composition and Doses. By a Practising Physician. Fifth Edition. 32mo.
cloth, 2s. 6d.; roan tuck (for the pocket), 3s. 6d.

DR. JOHN ROWLISON PRETTY.

## AIDS DURING LABOUR, including the Administration of Chloroform,
the Management of Placenta and Post-partum Hæmorrhage. Fcap. 8vo. cloth, 4s. 6d.

MR. P. C. PRICE, F.R.C.S.

## AN ESSAY ON EXCISION OF THE KNEE-JOINT. With
Coloured Plates. With Memoir of the Author and Notes by Henry Smith, F.R.C.S.
Royal 8vo. cloth, 14s.

**MR. LAKE PRICE.**

PHOTOGRAPHIC MANIPULATION : A Manual treating of the Practice of the Art, and its various Applications to Nature. With numerous Engravings. Second Edition. Crown 8vo. cloth, 6s. 6d.

**DR. PRIESTLEY.**

LECTURES ON THE DEVELOPMENT OF THE GRAVID UTERUS. 8vo. cloth, 5s. 6d.

**DR. RADCLIFFE, F.R.C.P.L.**

LECTURES ON EPILEPTIC, SPASMODIC, NEURALGIC, AND PARALYTIC DISORDERS OF THE NERVOUS SYSTEM, delivered at the Royal College of Physicians in London. Post 8vo. cloth, 7s. 6d.

**MR. RAINEY.**

ON THE MODE OF FORMATION OF SHELLS OF ANIMALS, OF BONE, AND OF SEVERAL OTHER STRUCTURES, by a Process of Molecular Coalescence, Demonstrable in certain Artificially-formed Products. Fcap. 8vo. cloth, 4s. 6d.

**DR. F. H. RAMSBOTHAM.**

THE PRINCIPLES AND PRACTICE OF OBSTETRIC MEDI-CINE AND SURGERY. Illustrated with One Hundred and Twenty Plates on Steel and Wood; forming one thick handsome volume. Fifth Edition. 8vo. cloth, 22s.

**DR. RAMSBOTHAM.**

PRACTICAL OBSERVATIONS ON MIDWIFERY, with a Selection of Cases. Second Edition. 8vo. cloth, 12s.

**DR. READE, M.B.T.C.D., L.R.C.S.I.**

SYPHILITIC AFFECTIONS OF THE NERVOUS SYSTEM, AND A CASE OF SYMMETRICAL MUSCULAR ATROPHY ; with other Contributions to the Pathology of the Spinal Marrow. Post 8vo. cloth, 5s.

**PROFESSOR REDWOOD, PH.D.**

A SUPPLEMENT TO THE PHARMACOPŒIA : A concise but comprehensive Dispensatory, and Manual of Facts and Formulæ, for the use of Practitioners in Medicine and Pharmacy. Third Edition. 8vo. cloth, 22s.

**DR. DU BOIS REYMOND·**

ANIMAL ELECTRICITY ; Edited by H. BENCE JONES, M.D., F.R.S. With Fifty Engravings on Wood. Foolscap 8vo. cloth, 6s.

**DR. REYNOLDS, M.D.LOND.**

I.

EPILEPSY: ITS SYMPTOMS, TREATMENT, AND RELATION TO OTHER CHRONIC CONVULSIVE DISEASES. 8vo. cloth, 10s.

II.

THE DIAGNOSIS OF DISEASES OF THE BRAIN, SPINAL CORD, AND THEIR APPENDAGES. 8vo. cloth, 8s.

**DR. B. W. RICHARDSON.**

ON THE CAUSE OF THE COAGULATION OF THE BLOOD. Being the ASTLEY COOPER PRIZE ESSAY for 1856. With a Practical Appendix. 8vo. cloth, 16s.

DR. RITCHIE, M.D.

## ON OVARIAN PHYSIOLOGY AND PATHOLOGY. With

Engravings. 8vo. cloth, 6s.

DR. WILLIAM ROBERTS, M.D., F.R.C.P.

## AN ESSAY ON WASTING PALSY; being a Systematic Treatise on

the Disease hitherto described as ATROPHIE MUSCULAIRE PROGRESSIVE.
With Four Plates. 8vo. cloth, 5s.

DR. ROUTH.

## INFANT FEEDING, AND ITS INFLUENCE ON LIFE;

Or, the Causes and Prevention of Infant Mortality. Second Edition. Fcap. 8vo. cloth, 6s.

DR. W. H. ROBERTSON.

I.

## THE NATURE AND TREATMENT OF GOUT. 8vo. cloth, 10s. 6d.

II.

## A TREATISE ON DIET AND REGIMEN. Fourth Edition. 2 vols.

12s. post 8vo. cloth.

DR. ROWE.

## NERVOUS DISEASES, LIVER AND STOMACH COM-

PLAINTS, LOW SPIRITS, INDIGESTION, GOUT, ASTHMA, AND DIS-
ORDERS PRODUCED BY TROPICAL CLIMATES. With Cases. Sixteenth
Edition. Fcap. 8vo. 2s. 6d.

DR. ROYLE, F.R.S., AND DR. HEADLAND, M.D.

## A MANUAL OF MATERIA MEDICA AND THERAPEUTICS.

With numerous Engravings on Wood. Fifth Edition. Fcap. 8vo. cloth, 12s. 6d.

DR. RYAN, M.D.

## INFANTICIDE: ITS LAW, PREVALENCE, PREVENTION, AND

HISTORY. 8vo. cloth, 5s.

ST. BARTHOLOMEW'S HOSPITAL.

## A DESCRIPTIVE CATALOGUE OF THE ANATOMICAL

MUSEUM. Vol. I. (1846), Vol. II. (1851), Vol. III. (1862), 8vo. cloth, 5s. each.

## ST. GEORGE'S HOSPITAL REPORTS. Vols. I. & II. 8vo. 7s. 6d.

MR. T. P. SALT, BIRMINGHAM.

I.

## ON DEFORMITIES AND DEBILITIES OF THE LOWER

EXTREMITIES, AND THE MECHANICAL TREATMENT EMPLOYED
IN THE PROMOTION OF THEIR CURE. With Plates. 8vo. cloth, 15s.

II.

## ON RUPTURE: ITS CAUSES, MANAGEMENT, AND CURE,

and the various Mechanical Contrivances employed for its Relief. With Engravings.
Post 8vo. cloth, 3s.

DR. SALTER, F.R.S.

## ASTHMA. Second Edition. 8vo. cloth, 10s.

DR. SANKEY, M.D.LOND.

## LECTURES ON MENTAL DISEASES. 8vo. cloth, 8s.

### DR. SANSOM, M.D.LOND.

I.

CHLOROFORM: ITS ACTION AND ADMINISTRATION. A Handbook. With Engravings. Crown 8vo. cloth, 5s.

II.

THE ARREST AND PREVENTION OF CHOLERA; being a Guide to the Antiseptic Treatment. Fcap. 8vo. cloth, 2s. 6d.

### MR. SAVORY.

A COMPENDIUM OF DOMESTIC MEDICINE, AND COMPANION TO THE MEDICINE CHEST; intended as a Source of Easy Reference for Clergymen, and for Families residing at a Distance from Professional Assistance. Seventh Edition. 12mo. cloth, 5s.

### DR. SCHACHT.

THE MICROSCOPE, AND ITS APPLICATION TO VEGETABLE ANATOMY AND PHYSIOLOGY. Edited by FREDERICK CURREY, M.A. Fcap. 8vo. cloth, 6s.

### DR. SCORESBY-JACKSON, M.D., F.R.S.E.

MEDICAL CLIMATOLOGY; or, a Topographical and Meteorological Description of the Localities resorted to in Winter and Summer by Invalids of various classes both at Home and Abroad. With an Isothermal Chart. Post 8vo. cloth, 12s.

### DR. SEMPLE.

ON COUGH: its Causes, Varieties, and Treatment. With some practical Remarks on the Use of the Stethoscope as an aid to Diagnosis. Post 8vo. cloth, 4s. 6d.

### DR. SEYMOUR.

I.

ILLUSTRATIONS OF SOME OF THE PRINCIPAL DISEASES OF THE OVARIA: their Symptoms and Treatment; to which are prefixed Observations on the Structure and Functions of those parts in the Human Being and in Animals. On India paper. . Folio, 16s.

II.

THE NATURE AND TREATMENT OF DROPSY; considered especially in reference to the Diseases of the Internal Organs of the Body, which most commonly produce it. 8vo. 5s.

### DR. SHAPTER, M.D., F.R.C.P.

THE CLIMATE OF THE SOUTH OF DEVON, AND ITS INFLUENCE UPON HEALTH. Second Edition, with Maps. 8vo. cloth, 10s. 6d.

### MR. SHAW, M.R.C.S.

THE MEDICAL REMEMBRANCER; OR, BOOK OF EMERGENCIES. Fifth Edition. Edited, with Additions, by JONATHAN HUTCHINSON, F.R.C.S. 32mo. cloth, 2s. 6d.

### DR. SHEA, M.D., B.A.

A MANUAL OF ANIMAL PHYSIOLOGY. With an Appendix of Questions for the B.A. London and other Examinations. With Engravings. Foolscap 8vo. cloth, 5s. 6d.

### DR. SHRIMPTON.

CHOLERA: ITS SEAT, NATURE, AND TREATMENT. With Engravings. 8vo. cloth, 4s. 6d.

### MR. U. J. KAY-SHUTTLEWORTH.

FIRST PRINCIPLES OF MODERN CHEMISTRY: a Manual of Inorganic Chemistry. Crown 8vo. cloth, 4s. 6d.

### DR. SIBSON, F.R.S.

MEDICAL ANATOMY. With coloured Plates. Imperial folio. Fasciculi I. to VI. 5s. each.

### DR. E. H. SIEVEKING.

ON EPILEPSY AND EPILEPTIFORM SEIZURES: their Causes, Pathology, and Treatment. Second Edition. Post 8vo. cloth, 10s. 6d.

### DR. SIMMS.

A WINTER IN PARIS : being a few Experiences and Observations of French Medical and Sanitary Matters. Fcap. 8vo. cloth, 4s.

### MR. SINCLAIR AND DR JOHNSTON.

PRACTICAL MIDWIFERY : Comprising an Account of 13,748 Deliveries, which occurred in the Dublin Lying-in Hospital, during a period of Seven Years. 8vo. cloth, 10s.

### DR. SIORDET, M.B.LOND., M.R.C.P.

MENTONE IN ITS MEDICAL ASPECT. Foolscap 8vo. cloth, 2s. 6d.

### MR. ALFRED SMEE, F.R.S.

GENERAL DEBILITY AND DEFECTIVE NUTRITION ; their Causes, Consequences, and Treatment. Second Edition. Fcap. 8vo. cloth, 3s. 6d.

### DR. SMELLIE.

OBSTETRIC PLATES: being a Selection from the more Important and Practical Illustrations contained in the Original Work. With Anatomical and Practical Directions. 8vo. cloth, 5s.

### MR. HENRY SMITH, F.R.C.S.

**I.**

ON STRICTURE OF THE URETHRA. 8vo. cloth, 7s. 6d.

**II.**

HÆMORRHOIDS AND PROLAPSUS OF THE RECTUM : Their Pathology and Treatment, with especial reference to the use of Nitric Acid. Third Edition. Fcap. 8vo. cloth, 3s.

**III.**

THE SURGERY OF THE RECTUM. Lettsomian Lectures. Second Edition. Fcap. 8vo. 3s. 6d.

### DR. J. SMITH, M.D., F.R.C.S.EDIN.

HANDBOOK OF DENTAL ANATOMY AND SURGERY, FOR THE USE OF STUDENTS AND PRACTITIONERS. Fcap. 8vo. cloth, 3s. 6d.

### DR. W. TYLER SMITH.

A MANUAL OF OBSTETRICS, THEORETICAL AND PRACTICAL. Illustrated with 186 Engravings. Fcap. 8vo. cloth, 12s. 6d.

### DR. SNOW.

ON CHLOROFORM AND OTHER ANÆSTHETICS: THEIR ACTION AND ADMINISTRATION. Edited, with a Memoir of the Author, by Benjamin W. Richardson, M.D. 8vo. cloth, 10s. 6d.

### MR. J. VOSE SOLOMON, F.R.C.S.

TENSION OF THE EYEBALL; GLAUCOMA : some Account of the Operations practised in the 19th Century. 8vo. cloth, 4s.

DR. STANHOPE TEMPLEMAN SPEER.

## PATHOLOGICAL CHEMISTRY, IN ITS APPLICATION TO
THE PRACTICE OF MEDICINE. Translated from the French of MM. BECQUEREL and RODIER. 8vo. cloth, reduced to 8s.

MR. J. K. SPENDER, M.B.LOND.

## A MANUAL OF THE PATHOLOGY AND TREATMENT
OF ULCERS AND CUTANEOUS DISEASES OF THE LOWER LIMBS. 8vo. cloth, 4s.

MR. BALMANNO SQUIRE, M.B.LOND.

I.

## CLINICAL LECTURES ON SKIN DISEASES. Illustrated by
Coloured Photographs from Life. Complete in 36 Numbers, price 1s. 6d. each. Nos. I.—XXX. are now ready.

II.

## A MANUAL OF THE DISEASES OF THE SKIN. Illustrated
by Coloured Plates of the Diseases, and by Woodcuts of the Parasites of the Skin. Post 8vo. cloth, 24s.

MR. PETER SQUIRE.

I.

## A COMPANION TO THE BRITISH PHARMACOPÆIA.
Sixth Edition. 8vo. cloth, 8s. 6d.    II.

## THE PHARMACOPÆIAS OF THIRTEEN OF THE LONDON
HOSPITALS, arranged in Groups for easy Reference and Comparison. 18mo. cloth, 3s. 6d.

DR. STEGGALL.

I.

## A MEDICAL MANUAL FOR APOTHECARIES' HALL AND OTHER MEDICAL
BOARDS. Twelfth Edition. 12mo. cloth, 10s.

II.

## A MANUAL FOR THE COLLEGE OF SURGEONS; intended for the Use
of Candidates for Examination and Practitioners. Second Edition. 12mo. cloth, 10s.

III.

## FIRST LINES FOR CHEMISTS AND DRUGGISTS PREPARING FOR EX-
AMINATION AT THE PHARMACEUTICAL SOCIETY. Second Edition. 18mo. cloth, 3s. 6d.

MR. STOWE, M.R.C.S.

## A TOXICOLOGICAL CHART, exhibiting at one view the Symptoms,
Treatment, and Mode of Detecting the various Poisons, Mineral, Vegetable, and Animal. To which are added, concise Directions for the Treatment of Suspended Animation. Twelfth Edition, revised. On Sheet, 2s.; mounted on Roller, 5s.

MR. FRANCIS SUTTON, F.C.S.

## A SYSTEMATIC HANDBOOK OF VOLUMETRIC ANALYSIS;
or, the Quantitative Estimation of Chemical Substances by Measure. With Engravings. Post 8vo. cloth, 7s. 6d.

DR. SWAYNE.

## OBSTETRIC APHORISMS FOR THE USE OF STUDENTS
COMMENCING MIDWIFERY PRACTICE. With Engravings on Wood. Fourth Edition. Fcap. 8vo. cloth, 3s. 6d.

## MR. TAMPLIN, F.R.C.S.E.

LATERAL CURVATURE OF THE SPINE: its Causes, Nature, and Treatment. 8vo. cloth, 4s.

## SIR ALEXANDER TAYLOR, M.D., F.R.S.E.

THE CLIMATE OF PAU; with a Description of the Watering Places of the Pyrenees, and of the Virtues of their respective Mineral Sources in Disease. Third Edition. Post 8vo. cloth, 7s.

## DR. ALFRED S. TAYLOR, F.R.S.

I.

THE PRINCIPLES AND PRACTICE OF MEDICAL JURIS-PRUDENCE. With 176 Wood Engravings. 8vo. cloth, 28s.

II.

A MANUAL OF MEDICAL JURISPRUDENCE. Eighth Edition. With Engravings. Fcap. 8vo. cloth, 12s. 6d.

III.

ON POISONS, in relation to MEDICAL JURISPRUDENCE AND MEDICINE. Second Edition. Fcap. 8vo. cloth, 12s. 6d.

## MR. TEALE.

ON AMPUTATION BY A LONG AND A SHORT RECTAN-GULAR FLAP. With Engravings on Wood. 8vo. cloth, 5s.

## DR. THEOPHILUS THOMPSON, F.R.S.

CLINICAL LECTURES ON PULMONARY CONSUMPTION; with additional Chapters by E. SYMES THOMPSON, M.D. With Plates. 8vo. cloth, 7s. 6d.

## DR. THOMAS.

THE MODERN PRACTICE OF PHYSIC; exhibiting the Symptoms, Causes, Morbid Appearances, and Treatment of the Diseases of all Climates. Eleventh Edition. Revised by ALGERNON FRAMPTON, M.D. 2 vols. 8vo. cloth, 28s.

## SIR HENRY THOMPSON, F.R.C.S.

I.

STRICTURE OF THE URETHRA; its Pathology and Treatment. The Jacksonian Prize Essay for 1852. With Plates. Second Edition. 8vo. cloth, 10s.

II.

THE DISEASES OF THE PROSTATE; their Pathology and Treatment. With Plates. Third Edition. 8vo. cloth, 10s.

III.

PRACTICAL LITHOTOMY AND LITHOTRITY; or, An Inquiry into the best Modes of removing Stone from the Bladder. With numerous Engravings. 8vo. cloth, 9s.

## DR. THUDICHUM.

I.

A TREATISE ON THE PATHOLOGY OF THE URINE, Including a complete Guide to its Analysis. With Plates, 8vo. cloth, 14s.

II.

A TREATISE ON GALL STONES: their Chemistry, Pathology, and Treatment. With Coloured Plates. 8vo. cloth, 10s.

### DR. TILT.

#### I.

## ON UTERINE AND OVARIAN INFLAMMATION, AND ON
THE PHYSIOLOGY AND DISEASES OF MENSTRUATION. Third Edition.
8vo. cloth, 12s.

#### II.

## A HANDBOOK OF UTERINE THERAPEUTICS, AND OF
MODERN PATHOLOGY OF DISEASES OF WOMEN. Second Edition.
Post 8vo. cloth, 6s.

#### III.

## THE CHANGE OF LIFE IN HEALTH AND DISEASE: a
Practical Treatise on the Nervous and other Affections incidental to Women at the Decline
of Life. Second Edition. 8vo. cloth, 6s.

### DR. GODWIN TIMMS.

## CONSUMPTION: its True Nature and Successful Treatment. Re-issue,
enlarged. Crown 8vo. cloth, 10s.

### DR. ROBERT B. TODD, F.R.S.

#### I.

## CLINICAL LECTURES ON THE PRACTICE OF MEDICINE.
*New Edition, in one Volume, Edited by* Dr. BEALE, 8vo. cloth, 18s.

#### II.

## ON CERTAIN DISEASES OF THE URINARY ORGANS, AND
ON DROPSIES. Fcap. 8vo. cloth, 6s.

### MR. TOMES, F.R.S.

## A MANUAL OF DENTAL SURGERY. With 208 Engravings on
Wood. Fcap. 8vo. cloth, 12s. 6d.

### DR. TUNSTALL, M.D., M.R.C.P.

## THE BATH WATERS: their Uses and Effects in the Cure and
Relief of various Chronic Diseases. Fourth Edition, revised. Crown 8vo. cloth, 2s.

### DR. TURNBULL.

#### I.

## AN INQUIRY INTO THE CURABILITY OF CONSUMPTION,
ITS PREVENTION, AND THE PROGRESS OF IMPROVEMENT IN THE
TREATMENT. Third Edition. 8vo. cloth, 6s.

#### II.

## A PRACTICAL TREATISE ON DISORDERS OF THE STOMACH
with FERMENTATION; and on the Causes and Treatment of Indigestion, &c. 8vo.
cloth, 6s.

### DR. TWEEDIE, F.R.S.

## CONTINUED FEVERS: THEIR DISTINCTIVE CHARACTERS,
PATHOLOGY, AND TREATMENT. With Coloured Plates. 8vo. cloth, 12s.

## VESTIGES OF THE NATURAL HISTORY OF CREATION.
Eleventh Edition. Illustrated with 106 Engravings on Wood. 8vo. cloth, 7s. 6d.

### DR. UNDERWOOD

# TREATISE ON THE DISEASES OF CHILDREN. Tenth Edition,
with Additions and Corrections by HENRY DAVIES, M.D. 8vo. cloth, 15s.

### DR. UNGER.

# BOTANICAL LETTERS. Translated by Dr. B. PAUL. Numerous
Woodcuts. Post 8vo., 2s. 6d.

### MR. WADE, F.R.C.S.

# STRICTURE OF THE URETHRA, ITS COMPLICATIONS
AND EFFECTS; a Practical Treatise on the Nature and Treatment of those
Affections. Fourth Edition. 8vo. cloth, 7s. 6d.

### DR. WALKER, M.B.LOND.

# ON DIPHTHERIA AND DIPHTHERITIC DISEASES. Fcap.
8vo. cloth, 3s.

### DR. WALLER.

# ELEMENTS OF PRACTICAL MIDWIFERY; or, Companion to
the Lying-in Room. Fourth Edition, with Plates. Fcap. cloth, 4s. 6d.

### MR. HAYNES WALTON, F.R.C.S.

# SURGICAL DISEASES OF THE EYE. With Engravings on
Wood. Second Edition. 8vo. cloth, 14s.

### DR. WARING, M.D., M.R.C.P.LOND.

I.
# A MANUAL OF PRACTICAL THERAPEUTICS. Second Edition,
Revised and Enlarged. Fcap. 8vo. cloth, 12s. 6d.

II.
# THE TROPICAL RESIDENT AT HOME. Letters addressed to
Europeans returning from India and the Colonies on Subjects connected with their Health
and General Welfare. Crown 8vo. cloth, 5s.

### DR. WATERS, F.R.C.P.

I.
# DISEASES OF THE CHEST. CONTRIBUTIONS TO THEIR
CLINICAL HISTORY, PATHOLOGY, AND TREATMENT. With Plates.
8vo. cloth, 12s. 6d. II.

# THE ANATOMY OF THE HUMAN LUNG. The Prize Essay
to which the Fothergillian Gold Medal was awarded by the Medical Society of London.
Post 8vo. cloth, 6s. 6d. III.

# RESEARCHES ON THE NATURE, PATHOLOGY, AND
TREATMENT OF EMPHYSEMA OF THE LUNGS, AND ITS RELA-
TIONS WITH OTHER DISEASES OF THE CHEST. With Engravings. 8vo.
cloth, 5s.

### DR. ALLAN WEBB, F.R.C.S.L.

# THE SURGEON'S READY RULES FOR OPERATIONS IN
SURGERY. Royal 8vo. cloth, 10s. 6d.

### DR. WEBER.

## A CLINICAL HAND-BOOK OF AUSCULTATION AND PER-
CUSSION. Translated by JOHN COCKLE, M.D. 5s.

### MR. SOELBERG WELLS, M.D., M.R.C.S.

## ON LONG, SHORT, AND WEAK SIGHT, and their Treatment by
the Scientific Use of Spectacles. Second Edition. With Plates. 8vo. cloth, 6s.

### MR. T. SPENCER WELLS, F.R.C.S.

I.

## DISEASES OF THE OVARIES: THEIR DIAGNOSIS AND
TREATMENT. Vol. I. 8vo. cloth, 9s.

II.

## SCALE OF MEDICINES FOR MERCHANT VESSELS.
With Observations on the Means of Preserving the Health of Seamen, &c. &c.
Seventh Thousand. Fcap. 8vo. cloth, 3s. 6d.

### DR. WEST.

## LECTURES ON THE DISEASES OF WOMEN. Third Edition.
8vo. cloth, 16s.

### DR. UVEDALE WEST.

## ILLUSTRATIONS OF PUERPERAL DISEASES. Second Edi-
tion, enlarged. Post 8vo. cloth, 5s.

### MR. WHEELER.

## HAND-BOOK OF ANATOMY FOR STUDENTS OF THE
FINE ARTS. With Engravings on Wood. Fcap. 8vo., 2s. 6d.

### DR. WHITEHEAD, F.R.C.S.

## ON THE TRANSMISSION FROM PARENT TO OFFSPRING
OF SOME FORMS OF DISEASE, AND OF MORBID TAINTS AND
TENDENCIES. Second Edition. 8vo. cloth, 10s. 6d.

### DR. WILLIAMS, F.R.S.

## PRINCIPLES OF MEDICINE : An Elementary View of the Causes,
Nature, Treatment, Diagnosis, and Prognosis, of Disease. With brief Remarks on
Hygienics, or the Preservation of Health. The Third Edition. 8vo. cloth, 15s.

### DR. CHARLES T. WILLIAMS, M.B.OXON.

## THE CLIMATE OF THE SOUTH OF FRANCE, AND ITS
VARIETIES, MOST SUITABLE FOR INVALIDS ; with Remarks on Italian
and other Winter Stations. Crown 8vo. cloth, 3s. 6d.

## THE WIFE'S DOMAIN : the YOUNG COUPLE—the MOTHER—the NURSE
—the NURSLING. Post 8vo. cloth, 3s. 6d.

### DR. WINSLOW, M.D., D.C.L.OXON.

## OBSCURE DISEASES OF THE BRAIN AND MIND.
Fourth Edition. Carefully Revised. Post 8vo. cloth, 10s. 6d.

### MR. ERASMUS WILSON, F.R.S.

I.

## THE ANATOMIST'S VADE-MECUM: A SYSTEM OF HUMAN
ANATOMY. With numerous Illustrations on Wood. Eighth Edition. Foolscap 8vo. cloth, 12s. 6d.

II.

## ON DISEASES OF THE SKIN: A SYSTEM OF CUTANEOUS
MEDICINE. Sixth Edition. 8vo. cloth, 18s.
THE SAME WORK; illustrated with finely executed Engravings on Steel, accurately coloured. 8vo. cloth, 36s.

III.

## HEALTHY SKIN: A Treatise on the Management of the Skin and Hair
in relation to Health. Seventh Edition. Foolscap 8vo. 2s. 6d.

IV.

## PORTRAITS OF DISEASES OF THE SKIN. Folio. Fasciculi I.
to XII., completing the Work. 20s. each. The Entire Work, half morocco, £13.

V.

## THE STUDENT'S BOOK OF CUTANEOUS MEDICINE AND
DISEASES OF THE SKIN. Post 8vo. cloth, 8s. 6d.

VI.

## ON SYPHILIS, CONSTITUTIONAL AND HEREDITARY;
AND ON SYPHILITIC ERUPTIONS. With Four Coloured Plates. 8vo. cloth, 16s.

VII.

## A THREE WEEKS' SCAMPER THROUGH THE SPAS OF
GERMANY AND BELGIUM, with an Appendix on the Nature and Uses of Mineral Waters. Post 8vo. cloth, 6s. 6d.

VIII.

## THE EASTERN OR TURKISH BATH: its History, Revival in
Britain, and Application to the Purposes of Health. Foolscap 8vo., 2s.

### DR. WISE, M.D., F.R.C.P.EDIN.

## REVIEW OF THE HISTORY OF MEDICINE AMONG
ASIATIC NATIONS. Two Vols. 8vo. cloth, 16s.

### DR. G. C. WITTSTEIN.

## PRACTICAL PHARMACEUTICAL CHEMISTRY: An Explanation
of Chemical and Pharmaceutical Processes, with the Methods of Testing the Purity of the Preparations, deduced from Original Experiments. Translated from the Second German Edition, by STEPHEN DARBY. 18mo. cloth, 6s.

### DR. HENRY G. WRIGHT.

I.

## UTERINE DISORDERS: their Constitutional Influence and Treatment.
8vo. cloth, 7s. 6d.

II.

## HEADACHES; their Causes and their Cure. Fourth Edition. Fcap. 8vo.
2s. 6d.

### DR. YEARSLEY, M.D., M.R.C.S.

I.

## DEAFNESS PRACTICALLY ILLUSTRATED; being an Exposition
as to the Causes and Treatment of Diseases of the Ear. Sixth Edition. 8vo. cloth, 6s.

II.

## ON THROAT AILMENTS, MORE ESPECIALLY IN THE
ENLARGED TONSIL AND ELONGATED UVULA. Eighth Edition. 8vo. cloth, 5s.

# CHURCHILL'S SERIES OF MANUALS.

### Fcap. 8vo. cloth, 12s. 6d. each.

"We here give Mr. Churchill public thanks for the positive benefit conferred on the Medical Profession, by the series of beautiful and cheap Manuals which bear his imprint."— *British and Foreign Medical Review.*

## AGGREGATE SALE, 150,000 COPIES.

**ANATOMY.** With numerous Engravings. Eighth Edition. By ERASMUS WILSON, F.R.C.S., F.R.S.

**BOTANY.** With numerous Engravings. By ROBERT BENTLEY, F.L.S., Professor of Botany, King's College, and to the Pharmaceutical Society.

**CHEMISTRY.** With numerous Engravings. Ninth Edition. By GEORGE FOWNES, F.R.S., H. BENCE JONES, M.D., F.R.S., and A. W. HOFMANN, F.R.S.

**DENTAL SURGERY.** With numerous Engravings. By JOHN TOMES, F.R.S.

**MATERIA MEDICA.** With numerous Engravings. Fifth Edition. By J. FORBES ROYLE, M.D., F.R.S., and F. W. HEADLAND, M.D., F.L.S.

**MEDICAL JURISPRUDENCE.** With numerous Engravings. Eighth Edition. By ALFRED SWAINE TAYLOR, M.D., F.R.S.

**PRACTICE OF MEDICINE.** Second Edition. By G. HILARO BARLOW, M.D., M.A.

**The MICROSCOPE and its REVELATIONS.** With numerous Plates and Engravings. Fourth Edition. By W. B. CARPENTER, M.D., F.R.S.

**NATURAL PHILOSOPHY.** With numerous Engravings. Sixth Edition. By CHARLES BROOKE, M.B., M.A., F.R.S. *Based on the Work of the late Dr. Golding Bird.*

**OBSTETRICS.** With numerous Engravings. By W. TYLER SMITH, M.D., F.R.C.P.

**OPHTHALMIC MEDICINE and SURGERY.** With coloured Plates and Engravings on Wood. Third Edition. By T. WHARTON JONES, F.R.C.S., F.R.S.

**PATHOLOGICAL ANATOMY.** With numerous Engravings. By C. HANDFIELD JONES, M.B., F.R.S., and E. H. SIEVEKING, M.D., F.R.C.P.

**PHYSIOLOGY.** With numerous Engravings. Fourth Edition. By WILLIAM B. CARPENTER, M.D., F.R.S.

**POISONS.** Second Edition. By ALFRED SWAINE TAYLOR, M.D., F.R.S.

**PRACTICAL ANATOMY.** With numerous Engravings. (10s. 6d.) By CHRISTOPHER HEATH, F.R.C.S.

**PRACTICAL SURGERY.** With numerous Engravings. Fourth Edition. By Sir WILLIAM FERGUSSON, Bart., F.R.C.S., F.R.S.

**THERAPEUTICS.** Second Edition. By E. J. Waring, M.D., M.R.C.P.

Printed by W. BLANCHARD & SONS, 62, Millbank Street, Westminster.

www.ingramcontent.com/pod-product-compliance
Lightning Source LLC
Chambersburg PA
CBHW021818190326
41518CB00007B/646